AN INTRODUCTION TO POLITICAL GEOGRAPHY

'This innovative book is an excellent introduction to contemporary political geography. Its threefold structure provides valuable routes into material covering both current theoretical debates and illuminating case studies. It is up-to-date in every sense and enables students to appreciate the discipline's approaches through accessible exemplars.'

Ron Johnston, *University of Bristol*

'This book is much more than a basic introduction to political geography. It provides a critical, versatile alternative to traditional state-centric narratives in political geography and will be a valuable resource not only for political geographers but also for students in such fields as political science and sociology.'

Anssi Paasi, *University of Oulu*

Questions of the interaction between politics and geography permeate much of contemporary life. In this broad-based introduction to contemporary political geography, the authors examine the relationship between politics and geography at a variety of levels and in a number of different contexts. By pushing back the boundaries of what is conventionally understood to constitute political geography, the book emphasises the relationships between power, politics and policy, space, place and territory in different geographical contexts.

An Introduction to Political Geography explores how power interacts with space, how place influences political identities and how policy creates and remoulds territory. In outlining the full breadth of contemporary political geography, covering a rich and diverse range of topics, it addresses not only traditional concerns such as state formation, geopolitics, electoral geography and nationalism but also newer themes, including the geographies of regulation and governance, public policy, the politics of place consumption, landscapes of power, identity politics and geographies of resistance.

This accessible text successfully combines discussion of cutting-edge conceptual debates with international case studies, numerous illustrations and explanatory boxes. *An Introduction to Political Geography* will be essential reading for political geographers as well as a valuable resource for students of related fields with an interest in politics and geography.

Martin Jones is Reader in Human Geography, **Rhys Jones** is Lecturer in Human Geography and **Michael Woods** is Senior Lecturer in Human Geography in the Institute of Geography and Earth Sciences at the University of Wales Aberystwyth. They all teach on the Master's degree, Space, Place and Politics.

AN INTRODUCTION TO POLITICAL GEOGRAPHY

Space, place and politics

Martin Jones, Rhys Jones and Michael Woods

Routledge
Taylor & Francis Group

LONDON AND NEW YORK

First published 2004
by Routledge
2 Park Square, Milton Park, Abingdon, Oxon OX14 4RN

Simultaneously published in the USA and Canada
by Routledge
270 Madison Ave, New York, NY 10016

Reprinted 2006, 2007 (twice)

Routledge is an imprint of the Taylor & Francis Group, an informa business

Designed and typeset in Garamond by Keystroke, Jacaranda Lodge, Wolverhampton
Printed and bound in Great Britain by Bell & Bain Ltd, Glasgow

British Library Cataloguing in Publication Data
A catalogue record for this book is available from the British Library

Library of Congress Cataloging in Publication Data
Jones, Martin
An introduction to political geography: space, place and politics / Martin Jones,
Rhys Jones, Michael Woods.
p. cm.
Includes bibliographical references and index.
1. Political geography. I. Jones, Rhys. II. Woods, Michael. III. Title.
JC319.J66 2004
320.1′2–dc22
2003021646

ISBN-10: 0–415–25076–5 (hbk)
ISBN-10: 0–415–25077–3 (pbk)
ISBN-13: 978-0-415-25076-4 (hbk)
ISBN-13: 978-0-415-25077-1 (pbk)

Contents

Acknowledgements

Like many projects, this book has had a long gestation period. Between us, it is the result of nearly twenty years' curiosity with the broad field of political geography. The book is developed from a number of undergraduate and postgraduate courses that we have taught at the University of Wales Aberystwyth since 1995 and we would like to thank our many students for their perseverance and enthusiasm.

An Introduction to Political Geography would not have been completed without the assistance of a number of individuals. We owe a huge debt to Andrew Mould, who commissioned the book way back in 2000. Andrew has been an enthusiastic editor, mixing a number of well needed on-the-account meals and drinks – progress meetings in disguise – with emails asking 'exactly when are you going to deliver'. The answer is now (at last). We are also grateful to the anonymous reviewers, whose comments have proved valuable in reworking the manuscript.

We would like to thank Ian Gulley and Anthony Smith at the University of Wales Aberystwyth for redrawing some of the figures and maps that appear in the book. The authors and publishers would like to thank the following for granting permission to reproduce images: Plate 3.1, photograph by Gillian Jones; Figure 4.1, *Geoforum*, Elsevier; Figure 4.3, Ivan Turok; Table 4.2, *Geoforum*, Elsevier; Plate 5.1, photograph by David Henry; Figure 6.1, University of North Carolina Press; Figure 7.1, Cambridge University Press; Plate 8.1, photograph by Paul Routledge; and Figure 8.4, Brookings Institution. Every effort has been made to contact copyright holders for their permission to reprint material in this book. The publishers would be grateful to hear from any copyright holder who is not here acknowledged and will undertake to rectify any errors or omissions in future editions.

We would also like to acknowledge our various teachers, tutors, supervisors, mentors and colleagues for putting us on the right path over the years, and thanks also go to our friends and family for support during this project.

Martin Jones (msj@aber.ac.uk), home page http://users.aber.ac.uk/msj/
Rhys Jones (raj@aber.ac.uk), home page http://users.aber.ac.uk/raj/
Mike Woods (m.woods@aber.ac.uk), home page http://users.aber.ac.uk/zzp/woodshome.htm

Power, space and 'political geography'

Sydney, September 2000

It is the night of Monday 25 September 2000, in the closing week of the Olympic Games in Sydney, Australia. In front of a record crowd the Australian athlete Cathy Freeman sprints clear to win gold in the women's 400 metre final. It is Australia's first Olympic gold medal in athletics since 1988, and the hundredth medal won by an Australian since the start of the modern Olympics in 1896. Momentarily exhausted, Freeman sits cross-legged on the track, hands over her eyes and mouth. Then, collecting a flag from the trackside, she sets off on a barefoot lap of honour, draped in her dual-sided flag – on one face the 'southern cross' standard of Australia, on the other the red, black and gold Aboriginal flag.

Cathy Freeman's moment of Olympic history is saturated with political geography. Most explicitly, there is the demonstration of Australian patriotism, reflecting the way in which sports events often provide a focal point for the articulation of national identity. Yet, with Freeman, a black Aboriginal woman and Aboriginal rights campaigner, the event assumed a deeper, more complex, symbolism. Freeman had been reprimanded on a previous occasion when she had celebrated with the Aboriginal flag. This time, however, there were no objections as she waved her dual Australian and Aboriginal ensign. In doing so Freeman served not just to reaffirm Australian national identity but contributed to its reinvention, turning the Olympic stadium into the stage for a seminal performance in the politics of race and identity in Australia.

Freeman's celebrations refocused attention on the brutal oppression of the Aboriginal people during the British colonisation of 'Australia' as part of an imperial geopolitical strategy. Moreover, the subjugation of the Aboriginal people depended on the application of political geographic knowledge about the exercise of power through the control of space. Colonial authorities imposed new administrative territories without regard for any existing geographical understandings of the land, obliterated Aboriginal place names and tribal homelands, and exiled Aboriginal communities to spatially controlled 'reservations'.

Freeman was not the first to use the Olympic Games to make a political statement. The tradition includes the 'black power' salutes given by African-American athletes at the 1968 games in Mexico City, and the boycotts of the Moscow and Los Angeles games as part of geopolitical posturing in the 1980s. Today the very process of bidding to host the Olympics is a geo-political exercise, with competitors lobbying to build alliances of voting nations with negotiations that often spill over into issues of international diplomacy.

For the host city the prize is a symbolic step towards recognition as a 'global city'. The price, however, is a reworking of the city's own internal political geography. At Sydney, as at all the games, the stadium, athletes' village and the associated infrastructure of the event formed a 'landscape of power' which symbolised the powerfulness of the coalition of politicians, business leaders and sports administrators that had brought the games to Sydney, and the powerlessness of those who found themselves displaced by the development. The preparations for the games revealed much about the balance of power in contemporary urban politics as networks of key actors were assembled, funds diverted from health and education programmes, and new public order legislation introduced. At the same time, the Olympics became a site of resistance by Aboriginal rights and anti-globalisation protesters who defied new laws prohibiting demonstrations, claiming space and transgressing the spatial order of the 'Olympic city' as they did so.

These diverse stories from the Sydney Olympics illustrate the breadth and diversity of contemporary political geography. Some are about nation building, others about cultural politics, yet others about urban development or about governance – but they are all of interest to political geographers. In this book we provide an introduction to contemporary political geography that captures a sense of the dynamism and diversity of the sub-discipline at the start of the twenty-first century. As such, this book is by nature wide-ranging, covering topics from the medieval state to the regulation of the capitalist economy, and from community participation in planning in Berlin to conflicts over the use of the Confederate flag in South Carolina. What unites these seemingly disparate examples is that they all involve the interaction of 'politics' – defined in its broadest sense – and 'geography', represented by place, territory or spatial variation. It is this intersection of 'politics' and 'geography' that forms the central concern of this book and that is the basis of our understanding of 'political geography'.

Defining political geography

Political geographers have taken a number of different approaches to defining the field of political geography. To some, political geography has been about the study of political territorial units, borders and administrative subdivisions (Alexander 1963; Goblet 1955). For others, political geography is the study of political processes, differing from political science only in the emphasis given to geographical influences and outcomes and in the application of spatial analysis techniques (Burnett and Taylor 1981; Kasperson and Minghi 1969). Both these definitions reflected the influence of wider theoretical approaches within geography as a whole – regional geography and spatial

science respectively – at particular moments in the historical evolution of political geography and have generally been superseded as the discipline has moved on. Still current, however, is a third approach which holds that political geography should be defined in terms of its key concepts, which the proponents of this approach generally identify as territory and the state (e.g. Cox 2002). This approach shares with the earlier two approaches the desire to identify the 'essence' of political geography such that a definitive classification can be made of what is and what isn't 'political geography'. Yet political geography as it is actually researched and taught is much messier than these essentialist definitions suggest. Think, for example, about the word 'politics'. Essentialist definitions of political geography have tended to conceive of politics in very formal terms, as being about the state, elections and international relations. But 'politics' also occurs in all kinds of other, less formal, everyday situations, many of which have a strong geographical dimension – issues about the use of public space by young people for skateboarding, for example, or about the symbolic significance of a landscape threatened with development. While essentialist definitions of political geography would exclude most of these topics, they have become an increasingly important focus of geographical research.

As such, a fourth approach has been taken by writers who have sought to define political geography in a much more open and inclusive manner. John Agnew, for example, defines political geography as simply 'the study of how politics is informed by geography' (2002a: 1; see also Agnew *et al.* 2003), while Joe Painter (1995) describes political geography as a 'discourse', or a body of knowledge that produces particular understandings about the world, characterised by internal debate, the evolutionary adoption of new ideas, and dynamic boundaries. As indicated above, the way in which political geography is conceived of in this book fits broadly within this last approach.

We define political geography as a cluster of work within the social sciences that engages with the multiple intersections of 'politics' and 'geography', where these two terms are imagined as triangular configurations (Figure 1.1). On one side is the triangle

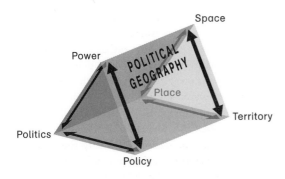

Figure 1.1 Political geography as the interaction of 'politics' and 'geography'

of power, politics and policy. Here power is the commodity that sustains the other two – as Bob Jessop puts it, 'if money makes the economic world go round, power is the medium of politics' (Jessop 1990a: 322) (see Box 1.1). Politics is the whole set of processes that are involved in achieving, exercising and resisting power – from the functions of the state to elections to warfare to office gossip. Policy is the intended outcome, the things that power allows one to achieve and that politics is about being in a position to do.

The interaction of these three entities is the concern of political science. Political geography is about the interaction of these entities and a second triangle of space, place and territory. In this triangle, space (or spatial patterns or spatial relations) is the core commodity of geography. Place is a particular point in space, while territory represents a more formal attempt to define and delimit a portion of space, inscribed with a particular identity and characteristics. Political geography recognises that these six entities – power, politics and policy, space, place and territory – are intrinsically linked, but a piece of political geographical research does not need to explicitly address them all. Spatial variations in policy implementation are a concern of political geography, as is the influence of territorial identity on voting behaviour, to pick two random examples. Political geography, therefore, embraces an innumerable multitude of interactions, some of which may have a cultural dimension which makes them also of interest to cultural geographers, some of which may have an economic dimension also of interest to economic geographers, some of which occurred in the past and are also studied by historical

---

BOX 1.1 POWER

Put simply, power is the ability to get things done, yet there are many different theories about what precisely power is and how it works. In broad terms there are two main approaches to conceptualising power. The first defines power as a property that can be possessed, building on an intellectual tradition that stems from Thomas Hobbes and Max Weber. Some writers in this tradition suggest that power is relational and involves conscious decision making, as Robert Dahl describes: 'A has power over B to the extent that he can get B to do something that B would not otherwise do' (Lukes 1974: 11–12). Others have argued that power can be possessed without being exercised, or that the exercise of power does not need conscious decision making but that ensuring that certain courses of action are never even considered is also an exercise of power. The second approach contends that power is not something that can be possessed, as Bruno Latour remarks: 'When you simply have power – *in potentia* – nothing happens and you are powerless; when you exert power – *in actu* – others are performing the action and not you. . . . History is full of people who, because they believed social scientists and deemed power to be something you can possess and capitalise, gave orders no one obeyed!' (Latour 1986: 264–5). Instead, power is conceived of as a 'capacity to act' which exists only when it is exercised and which requires the pooling together of the resources of a number of different entities.

Key readings: Clegg (1989) and Lukes (1986).

geographers. To employ a metaphor that we will explain in Chapter 2, political geography has frontier zones, not borders.

In this book we explore these various themes and topics by drawing on and discussing contemporary research in political geography. Nearly all the case studies and examples that we refer to are taken from books and journal articles published in the last twenty years, including many published since 2000 which may be regarded as at the 'cutting edge' of political geography research. However, current and recent work in political geography of this kind does not exist in a historical vacuum. It builds on the foundations of earlier research and writing, advancing an argument through critique and debate and through the exploration of new empirical studies that allow new ideas to be proposed. Knowing something about this genealogy of political geography helps us to understand the nature, approach and key concerns of contemporary political geography. To provide this background, the remainder of this chapter outlines a brief history of political geography, from the emergence of the sub-discipline in the nineteenth century to current debates about its future direction.

A brief history of political geography

The history of political geography as an academic sub-discipline can be roughly divided into three eras: an era of ascendancy from the late nineteenth century to the Second World War; an era of marginalisation from the 1940s to the 1970s; and an era of revival from the late 1970s onwards. However, the trajectory of political geographic writing and thinking can be traced back long before even the earliest of these dates. Aristotle, writing some 2,300 years ago in ancient Greece, produced a study of the state in which he adopted an environmental deterministic approach to considering the requirements for boundaries, the capital city, and the ratio between territory size and population; while the Greco-Roman geographer Strabo examined how the Roman Empire was able to overcome the difficulties caused by its great size to function

effectively. Interest in the factors shaping the form of political territories was revived in the European 'Age of Enlightenment' from the sixteenth century to the eighteenth, as writers combined their new enthusiasm for science and philosophy with the practical concerns generated by a period of political reform and instability. Most notable was Sir William Petty, an English scientist and economist who in 1672 published *The Political Anatomy of Ireland* in which he explored the territorial and demographic bases of the power of the British state in Ireland. Petty developed these ideas further in his second book, *Essays in Political Arithmetick*, begun in 1671 and published posthumously, which outlined theories on, among other things, a state's sphere of influence, the role of capital cities, and the importance of distance in limiting the reach of human activity. In this way Petty foreshadowed the concerns of many later political geographers, but, like other geographical writing of the time and the classical texts of Strabo and Aristotle, his books were popular works of individual scholarship by polymaths which did not stand as part of a coherent field of 'political geography'. To find the real beginnings of 'political geography' as an academic discipline we need to look to nineteenth-century Germany.

The era of ascendancy

The significance of Germany as the cradle of political geography lies in its relatively recent formation. Modern Germany had come into being as a unified state only in 1871 and under ambitious Prussian leadership sought in the closing decades of the nineteenth century to establish itself as a 'great power' on a par with Britain, France, Austria-Hungary and Russia. However, Germany was constrained by its largely landlocked, Central European location which restricted its potential for territorial expansion. In these circumstances, ideas about the relationship between territory and state power became key concerns for Germany's new intellectual class and, in particular, for Friedrich Ratzel, sometimes referred to as 'the father of political geography'.

Much of Ratzel's work was driven by a desire to justify intellectually the territorial expansion of

Germany, and in writings such as *Politische Geographie* he embarked on a 'scientific' study of the state (see Bassin 1987). Ratzel drew on earlier political geographical work, notably that of Carl Ritter, but his innovation was to borrow concepts from the evolutionary theories of Darwin and his followers. In particular, Ratzel was influenced by a variation on Social Darwinism known as neo-Lamarckism, which held that evolution occurred through species being directly modified by their environments rather than by chance. Translating these ideas to the political sphere, Ratzel argued that the state could be conceived of as a 'living organism' and that like every living organism the state 'required a specific amount of territory from which to draw sustenance. [Ratzel] labelled this territory the respective *Lebensraum* or living space of the particular organism' (Bassin 1987: 477).

Extending the metaphor, Ratzel contended that states followed the same laws of development as biological units and that when a state's *Lebensraum* became insufficient – for example, because of population growth – the state needed to annex new territory to establish new, larger, *Lebensraum*. As such he posited seven laws for the spatial growth of states, which held that a state must expand by annexing smaller territories, that in expanding a state strives to gain politically valuable positions, and that territorial expansion is contagious, spreading from state to state and intensifying, such that escalation towards warfare becomes inevitable. In this way Ratzel not only provided an 'intellectual justification' for German expansionism, but suggested that it was an entirely natural and necessary process. Ratzel himself argued that the only way Germany could acquire additional *Lebensraum* was through colonial expansion in Africa – a policy he actively promoted – but his theories were seen by some more militant nationalists as justifying the more aggressive and more dangerous strategy of expanding German territory in the crowded space of continental Europe itself.

Ratzel's ideas were developed further by Rudolf Kjellen, a Swedish conservative whose own political motives were fired by opposition to Norwegian independence. Kjellen's intellectual project was to develop a classification of states based on the Linnaean system.

By adapting Ratzel's theories, he attempted to identify the 'world powers' and predicted a future dominated by large continental imperialist states. Although he received some support in Germany, Kjellen's work would probably have been long forgotten had he not in an 1899 article coined the term *geopolitisk* which – translated into German as *Geopolitik* and by 1924 into English as *geopolitics* – came to describe that part of political geography that is essentially concerned with the external relations, strategy and politics of the state, and which seeks to employ such knowledge to political ends (see Chapter 3).

While Ratzel and Kjellen were wrestling with the dynamics of state power and territoriality, a second strand of political geography was being developed in Britain by Sir Halford Mackinder. Like Ratzel, Mackinder is regarded as a founding father of modern Geography, having popularised the subject in a series of public lectures in the 1880s and 1890s leading to his appointment as Oxford University's first Professor of Geography. Also like Ratzel, Mackinder saw the benefits of proving the political usefulness of his infant discipline. As O'Tuathail (1996: 25) has commented,

> to an ambitious intellectual like Mackinder, the governmentalizing of geographical discourse so that it addressed the imperialist dilemmas faced by Britain in a post-scramble world order was a splendid way of demonstrating the relevance of his 'new geography' to the ruling elites of the state.

However, unlike Ratzel, Mackinder was primarily concerned with issues of global strategy and the balance of power between states – topics that better suited the interests of British foreign policy. He was not the first to consider such matters. In the United States a retired naval officer, Alfred Mahan, had established himself as a newspaper pundit by arguing that global military power was dependent on sea power, and expounding on the geographical factors that enabled the development of a state as a sea power. Mackinder, though, disagreed with Mahan's thesis, suggesting that, as the age of exploration came to end, so the balance of power was shifting. In 1904 Mackinder published a paper entitled 'The geographical pivot of history' in the *Geographical Journal*, in which he divided history into three eras – a pre-Columbian era in which land power had been all-important, a Columbian era in which sea power had become predominant and an emergent post-Columbian era. In this new era, Mackinder argued, the end of the imperialist scramble had demoted the importance of sea power while new technologies which enabled long distances on land to be more easily overcome – such as the railways – would help to swing the balance of power back to continental states. Applying this hypothesis, Mackinder ordered the world map into three political regions – an 'outer crescent' across the Americas, Africa and the oceans; an 'inner crescent' across Europe and southern Asia; and the 'pivot area' located at the heart of the Eurasian land mass. Whoever controlled the pivot area, Mackinder argued, would be a major world power.

The First World War put the theories produced by the new political geography to the test, and Mackinder clearly felt that his ideas were vindicated. Writing in his 1919 book *Democratic Ideals and Reality*, he dismissed Ratzel's models as misguided and outdated:

> Last century, under the spell of the Darwinian theory, men came to think that those forms of organisation should survive which adapted themselves best to their natural environment. To-day we realise, as we emerge from our fiery trial, that human victory consists in our rising superior to such mere fatalism.
>
> (Mackinder 1919: 3)

In *Democratic Ideals and Reality* Mackinder expanded on his thesis of the shift from sea power to land power and recast his map of the world's seats of power to suit the new post-war order. He renamed the 'pivot area' the 'heartland', but left it centred on the Eurasian land mass, which he labelled the 'world island'. Significantly, he proposed that control of Eastern Europe was crucial to control of the heartland – and hence to global dominance (see Chapter 3). To maintain peace, therefore, Mackinder argued, Western Europe had to form a counterweight to Russia, which occupied the heartland, and the key priority of the West's strategy had to be to prevent Germany and

Russia forming an alliance that would dominate Eastern Europe.

Mackinder's ideas had a strong influence on the Versailles peace conference in 1919, in which he participated as a British delegate. Arguably, his legacy can be seen in the creation of 'buffer states' in Eastern Europe, separating Germany and Russia, more or less on the model that he proposes in *Democratic Ideals and Reality*. However, his continuing influence extended further than the map of Europe, informing US strategy in the Cold War, with the rhetoric and presumptions of Mackinder's heartland thesis surviving into the 1980s (see O'Tuathail 1992). Yet Mackinder was also criticised for oversimplifying history, underestimating the potential of air power and marginalising the significance of North America – a mistake which O'Tuathail (1992) describes as Mackinder's 'greatest blunder'. From this critique a modified approach was developed by writers such as Spykman (1942, 1944), which emphasised the strategic importance of the 'rimland' (or Mackinder's 'inner crescent') and which, by becoming closely related to US foreign policy, shifted the academic home of such theorising away from mainstream geography to international relations and strategic studies.

Ironically, Mackinder's thesis was also consumed with interest in the country that suffered most from its practical application at Versailles – Germany. For German nationalists, enraged by the way in which Germany had had its territory reduced and its military dismantled after the First World War, the geopolitical ideas of Ratzel and Mackinder offered a blueprint for revival (Paterson 1987). Most prominent in this movement was Karl Haushofer, a former military officer and geographer who became an early member of the Nazi party. Haushofer sought to build public support for a new expansionist policy by popularising interest in geopolitics. In 1924 he founded the *Zeitschrift für Geopolitik* (Journal of Geopolitics) and the following year was involved in establishing the German Academy, aimed at 'nourishing all spiritual expressions of Germandom', of which he later became president. Haushofer's 'pseudo-science' of *Geopolitik* took from Ratzel the concept of *Lebensraum* and twisted it, arguing that densely populated Germany needed to annex additional territory from more sparsely populated countries like Poland and Czechoslovakia. From Mackinder it took the idea that control of Eastern Europe and the heartland would lead to global dominance, arguing for the construction of a continental bloc comprising Germany, Russia and Japan which would control the heartland and form a counterweight to the British Empire (see O'Tuathail 1996).

Geopolitik provided the intellectual justification for Nazi Germany's annexation of Czechoslovakia and Poland, for the Hitler–Stalin pact and, later, for Germany's ill-fated invasion of the Soviet Union. However, the extent of Haushofer's influence on the Nazi leadership is questionable (see Heske 1987). More significant was the contribution of *Geopolitik* in shaping public opinion, most effectively achieved through the promotion of a new form of cartography in which highly subjective maps were used to emphasise the mismatch between Germany's post-1919 borders and its 'cultural sphere', to justify the annexation of territory and to suggest that it was vulnerable to aggression by its Slavic neighbours (see Herb 1997 for examples). The misadventures of *Geopolitik* inextricably associated political geography with the brutality and racism of the Nazi regime and led to its discrediting as a serious academic pursuit.

The era of marginalisation

The excesses of German *Geopolitik* cast a pall over all political geography. Writing in 1954, the leading American geographer Richard Hartshorne mournfully remarked of political geography that 'in perhaps no other branch of geography has the attempt to teach others gone so far ahead of the pursuit of learning by the teachers' (Hartshorne 1954: 178). In an attempt to 'depoliticise' political geography and to put it on what he regarded as a more scientific footing, Hartshorne (1950) promoted a 'functional approach' to political geography. He argued that political geography should be concerned not with shaping political strategy, but rather with describing and analysing the internal dynamics and external functions of the state. Included in the former were the centrifugal forces that placed pressures on the cohesion of states (such

as communication problems and ethnic differences), the centripetal forces which held states together (such as the state idea and the concept of a 'nation'), and the internal organisational mechanisms through which a state governed its territory. The external functions, meanwhile, included the territorial, economic, diplomatic and strategic relations of a state with other states.

The functional approach led political geographers to become concerned with questions such as the distribution of different ethnic populations in a state, the match between a state's boundaries and physical geographical features, and the structure of a state's local government areas, as well as with mapping patterns of communication networks within states and of trade routes between states. (Some examples of this type of work include Cole 1959; East and Moodie 1956; Moodie 1949; Soja 1968; Weigert 1949.) However, while the functional approach was popularised after the Second World War, it was pioneered in Britain and North America between the wars and arguably can be traced back to the work of Isaiah Bowman in the early 1920s.

Like Mackinder Bowman had been a participant in the Versailles talks, but unlike Mackinder he regarded the new world map that emerged as extremely unstable. His pessimism stemmed from concern not with strategic models, but with social and economic factors such as access to natural resources and the distribution of population, which he considered to be the real sources of political instability. Bowman set out these concerns in *The New World* (1921), in which he identified the 'major problems' facing the new world order as national debts and reparations, control over the production and distribution of raw materials, population movement and the distribution of land, the status of mandates and colonies, trade barriers and control over communications and transit links, the limitation of armaments, the status of minority populations and disputed boundaries between states. Bowman changed the scale at which political geography was focused and set the foundations for a new, arguably more scientific and more objective, form of analysis. This new style of political geography was more explicitly outlined by East (1937) in a paper which Johnston (1981) identifies as laying down the principles of the functional approach later championed by Harteshorne. East argued that 'the proper function of political geography is the study of the geographical results of political differentiation' and 'that the visible landscape is modified by the results of state and inter-state activities is a matter of common observation and experience' (East 1937: 263). As such, East continued,

> political geography is distinguishable from other branches of geography only in its subject matter and specific objectives. . . . Whereas the regional geographer has for his objective the discovery and description of the distinct components of a physical and human landscape . . . the political geographer analyses geographically the human and physical texture of political territories.
>
> (East 1937: 267)

Political geography as practised in the immediate post-Second World War period therefore had little by way of a distinct identity separate from mainstream regional geography, and became largely fixated on the territorial state as its object of analysis. Moreover, fear of the sub-discipline's past made political geographers wary of modelling and theorising, such that research remained essentially descriptive and empirically driven. The consequences of this self-restraint were twofold. First, political geography largely missed out on theoretical developments taking place elsewhere in geography, notably the 'quantitative revolution' of the late 1960s. Second, (and relatedly), political geography became marginalised within geography and began to disappear as a university subject. Berry (1969: 450) famously described it as 'that moribund backwater' and by the mid-1970s Muir (1976) found that political geography was taught in only half of Britain's university geography departments, with over two-thirds of heads of geography departments considering that the development of political geography literature was unsatisfactory compared with other branches of geography.

However, Muir's article, which was provocatively entitled 'Political geography – dead duck or phoenix?', found grounds for optimism. He noted that over half of respondents to his survey had felt that political geography was 'an underdeveloped branch of geography that *should* increase in importance' (Muir 1976: 196), and pointed to theoretical innovations that were beginning to take place on the fringes of the sub-discipline. He concluded, 'the contemporary climate of geographical opinion augers well for the future of political geography, and a promising trickle of progressive contributions suggests stimulating times to come' (p. 200).

The era of revival

The revival of political geography that Muir detected in the 1970s was driven by two parallel processes – the reintroduction of theory into political geography and a 'political turn' in geography more broadly. Significantly, neither resulted from developments in the established mainstream of political geography, but rather reflected innovation at the fringes of political geography, producing research clusters which eventually came to eclipse the old-style 'functional approach'. One illustration of this is the rise of quantitative electoral geography from the late 1960s onwards. Although the quantitative revolution tended to pass political geographers by, some quantitative geographers realised that the spatially structured nature of elections, combined with the large amount of easily available electoral data, made them an ideal focus for the application of quantitative geographical analysis. Elections had not traditionally been a concern of mainstream political geographers, and the new electoral geographers did not therefore have to challenge any orthodoxies as they employed quantitative techniques to develop models and test hypotheses across their tripartite interests of geographies of voting, geographical influences on voting and geographical analyses of electoral districts (Busteed 1975; McPhail 1971; Taylor and Johnston 1979). The lure of technical and theoretical innovation made electoral geography the fashionable 'cutting edge' of political geography in the 1970s, such that by 1981 Muir was moved to comment that its output had become 'disproportionate in relation to the general needs of political geography'

(Muir 1981: 204). (We discuss electoral geography in fuller detail in Chapter 8.)

The growth of electoral geography was the most prominent aspect of the belated introduction of a systems approach to political geography, drawing on the broader development of systems theory in geography as part of a focus on processes, not places (Cohen and Rosenthal 1971; Dikshit 1977). Electoral geographers viewed the electoral process as a system – comprised of various interacting parts, following certain rules and having particular spatial outcomes – but they also realised that other parts of the political world could also be conceived of and analysed as systems, including the state, local government, policy making and public spending (see Johnston 1979). Significantly, the mechanical principles underlying systems theory meant that adopting the approach rendered complex political entities suitable for mathematical analysis and modelling. However, the extent to which a full-bodied systems analysis was adopted in political geography varied. At the most basic level, 'systematic political geography' implied no more than reordering the way in which political geography was taught and researched to start from themes or concepts rather than regions (see de Bilj 1967). While this allowed generalisation in a way that the regionally focused approach did not, it did not necessarily lead to in-depth theorising. Yet even the most conscientious attempts to produce models and theories through quantitative analysis were constrained by their positivist epistemology – that is, the belief that the world might be understood through the construction and testing of laws based on empirical observation. As critics pointed out, positivism is problematic because it creates a false sense of objectivity, filters out social and ethical questions, oversimplifies the relation between observed events and theoretical languages, and fails to engage with the part played by both human agency and social, economic and political structures in shaping the human world (see Cloke *et al.* 1991; Gregory 2000b). Thus, because of these epistemological shortcomings, positivist political geography continued to be strangely apolitical (Johnston 1980). Moreover, the 'time lag' that afflicted the introduction of concepts into political geography meant that positivism was being championed in political geography at a time when these criticisms were already widely accepted elsewhere (Walsh 1979).

Ironically, the challenge to positivism was led by theoretical approaches that were intrinsically political, not least the development of Marxist political economy within geography (see Box 4.1 for more on models of political economy). In *Social Justice and the City* (1973), for example, David Harvey proposed a new analysis of urban systems as embedded in capitalism which described an urban geography saturated by class, corporate and state power and forged through political conflict. However, the infusion of these ideas into political geography was slow. Despite the calls of commentators such as Walsh that 'what political geography needs most urgently . . . is a comprehensive analysis of the state as a political-economic entity' (Walsh 1979: 92), political-economic research within political geography remained the exception, not the rule, and the task of studying urban conflicts, the geography of the state and the political–geographic expressions of capitalism was taken up primarily by urban and economic geographers, political scientists and sociologists. It was not until the 1980s that mainstream political geography really started to take the political-economy approach seriously, with the blossoming of work on the state, localities and urban politics (see Johnston 1989). The development of the political economy approach in political geography and its continuing in current research concern with state strategy, governance and the policy process is discussed in Chapter 4, while political economic approaches to local politics are among those discussed in Chapter 6.

One of the relatively few attempts to link the traditional concerns of political geography with theoretical insight from Marxist political economy was Peter Taylor's introduction of world systems analysis. The world systems approach had been developed by a political sociologist, Immanuel Wallerstein, who was himself influenced by the materialist school of historical analysis associated with Fernand Braudel and Karl Polányi and by neo-Marxist development studies (see Wallerstein 1979, 1991). As Box 1.2 details, Wallerstein rejected the idea that societal change could be studied on a country-by-country basis and argued instead that change at any scale can be understood only in the context of a 'world system'. The modern world system, Wallerstein argues, is global in scope, but he recognises that it is only the latest of a series of historical systems and proposes that it is the changes within and between historical systems that are the key to understanding contemporary society, economy and politics. For Taylor, the world systems approach was particularly attractive to political geography not only because spatial pattern was core to its analysis (Taylor 1988) but also because it offered the potential to develop a comprehensive, unifying theory of political geography that could include traditional areas like geopolitics and electoral geography and accommodate political-economic analysis of the state, urban politics and so on. However, despite its superficial attractiveness, world systems analysis is open to a number of criticisms (Box 1.2), and although it has formed the framework of Taylor's series of textbooks (see Taylor 1985 and Taylor and Flint 2000 as the first and most recent editions), the world systems approach has not been widely adopted by political geographers.

Far more influential have been two conceptual developments which served to further politicise the outlook of human geography as a whole. The first of these was the so-called 'cultural turn' of the late 1980s and 1990s which promoted a new understanding of culture as the product of discourses through which people signify their identity and experiences and which are constantly contested and renegotiated (see Jackson 1989; Mitchell 2000). Consequently, issues of power and resistance were positioned as central to the analysis of cultural geographies, generating significant clusters of research on questions of identity and place, including national identity and citizenship; conflict and contestation between cultural discourses; geographies of resistance; the role of landscape in conveying and challenging power; and 'micro-geographies' of politics, including investigation of the body as a site of oppression and resistance (see for example Pile and Keith 1997; Sharp et al. 2000). These themes are discussed further in Chapters 5, 7 and 8.

Moreover, the 'new cultural geography' drew on the conceptual writings of post-structuralist thinkers such

BOX 1.2 PETER TAYLOR, IMMANUEL WALLERSTEIN AND WORLD SYSTEMS ANALYSIS

World systems analysis forms the basis of the best-known attempt to construct a comprehensive theoretical framework for political geography, undertaken by Peter Taylor. It was initially developed by Immanuel Wallerstein as a critique of analyses of social change that focused on one country and considered only a short-term perspective. In contrast, two of the fundamental principles of world systems analysis are that social change at any scale can be understood only in the context of a wider world system, and that change needs to be approached through a long-term historical perspective. (The latter principle is derived from economic historians such as Fernand Braudel and Karl Polányi.)

Wallerstein holds that a single modern world system is now globally dominant, but that it has been preceded by numerous historical systems. These systems can be categorised as one of three types of 'entity', characterised by their mode of production. In the most basic, the mini-system, production is based on hunting, gathering or rudimentary agriculture where there is limited specialisation of tasks and exchange is reciprocal between producers. In the second type, the world empire, agricultural production creates a surplus that can support the expansion of non-agricultural production and the establishment of a military-bureaucratic elite. The third type, the world economy, is based on the capitalist mode of production where the aim of production is to create profit. From the sixteenth century onwards, Wallerstein argues, the European 'world economy' system expanded to subjugate all other systems and monopolise the globe. Transformation from one system to another can occur as a result of either internal or external factors, but changes can also occur within systems (termed 'continuities') – for example, in cycles of economic growth and stagnation. In the modern world economy these cycles are mapped by the Kondratieff waves which describe fifty-year cycles of growth and stagnation in the global economy since 1780/90.

Wallerstein further described the modern world economy as being defined by three basic elements. First, there is a single world market, which is capitalist, and in which competition results in uneven economic development across the world. Second, there is a multiple state system. The existence of different states is seen as a necessary condition for economic competition, but it also results in political competition between states, creating a variety of 'balances of power' over time. Third, the world economy always operates in a three-tier format. As Taylor and Flint (2000) explain: 'in any situation of inequality three tiers of interaction are more stable than two tiers of confrontation. Those at the top will always manoeuvre for the 'creation' of a three-tier structure, whereas those at the bottom will emphasize the two tiers of 'them and us'. The continuing existence of the world-economy is therefore due in part to the success of the ruling groups in sustaining three-tier patterns throughout various fields of conflict' (p. 12). Examples cited by Taylor and Flint include 'centre' parties in democratic political systems and the 'middle class', but also, crucially, a geographical ordering of the world into 'core', 'periphery' and 'semi-periphery'. For Wallerstein, core areas are associated with complex production regimes, and the periphery with more rudimentary structures. But there is also a 'semi-periphery' in which elements of both core and peripheral processes can be found, and which forms a dynamic zone where opportunities for political and economic change exist.

By drawing on these different components of world systems theory, Taylor identified a 'space–time matrix' for political geography, structured by Kondratieff cycle and spatial position (core, periphery or semi-periphery), which formed a context for the analysis of all types of political interaction from the global scale down to the household scale, hence providing a unifying framework for political geography.

However, the world systems approach can be criticised on a number of grounds. First, it is economically reductionist – it sees the driving processes of change as purely economic; it positions political action as secondary; and it reduces sexism and racism to reflections of the economy. Second, it is totalising in that it incorporates everything under one big umbrella and fails to acknowledge fully the heterogeneity of political or cultural relations. Third, it is functionalist, not recognising that what causes something to exist may have nothing to do with the effects it produces. For example, the factors behind the creation of a nation-state may not be related to subsequent nationalist actions.

Key readings: For more on world systems analysis see Taylor and Flint (2000), especially chapter 1, and Wallerstein (1991). For more on the critique of world systems analysis see Painter (1995) and Giddens (1985).

as Michel Foucault, Jacques Derrida, Gilles Deleuze and Félix Guattari, and postcolonial theorists such as Homi Bhabha, for whom the relation of power and space was a key concern (see Box 1.3). A number of different strands of post-structuralist thought have been introduced into political geography, including ideas about difference in research on the cultural politics of identity and the use of Derrida's method of deconstruction in critical geopolitics (see below). However, it is the work of Michel Foucault that has arguably had the greatest influence in political geography, in particular through the development and application of two key concepts. The first of these is 'discourse', which Foucault redefined as referring to the ensemble of social practices through which the world is made meaningful but which are also dynamic and contested (Box 1.4). In books such as *The Order of Things* (1973 [1966]) and *The Archaeology of Knowledge* (1974 [1969]) Foucault examined the articulation of discursive practices and thus established precedents as to how discourses might be analysed. These ideas have been fundamental to the development of geographical work on cultural politics and of critical geopolitics, as well as to the development of discourse analysis as a methodological approach which is now widely used across political geography. The second key concept is 'governmentality', by which Foucault refers to the means by which government renders society governable. Governmentality is essentially about the use of particular 'apparatuses of knowledge' and has been employed in recent years in work on the state and citizenship (see Chapter 8).

A significant aspect of both discourse analysis and governmentality is the potential they allow for exploration of the incorporation of space itself as a tool in the exercise of power. Much of Foucault's writing was concerned with power, but he rejected conventional notions of power as a property that is possessed, focusing instead on how power is exercised and how it circulates through society. Foucault stated that 'space is fundamental in any exercise of power' (Rabinow 1984: 252), and this principle underlies much of his work on disciplinary power. His best known illustration of this is his discussion of Jeremy Bentham's panopticon (Foucault 1977: ch. 3). The panopticon was a proposal for an ideal prison, the spatial arrangement of which would effectively force prisoners to discipline themselves. The panopticon would be built in a circular

BOX 1.3 POST-STRUCTURALISM

Post-structuralism refers to the theories advanced by a loose collection of philosophical writing produced in the late twentieth century, most notably in France. Labelled 'post-structuralism' because of the way it built on earlier structuralist theories, the approach is particularly identified with the work of Jacques Derrida, Michel Foucault, Gilles Deleuze, Julia Kristeva and Jean Baudrillard. The core ideas of post-structuralism are the rejection of the notion of an essential 'truth' and the consequential examination of the notion of 'difference'. Building on the work of structuralist thinkers like Saussure (1983), post-structuralists hold that language does not reflect meaning, but rather that meaning is produced within language and that the relation between the signifier (a sound or written image) and the signified (the meaning) is never fixed. Moreover, post-structuralists reject the idea of the rational subject, arguing that subjectivity (the sense of who we are) is constructed through discourses (see Box 1.4) that are open to change and contestation, and that there is no external 'reality' outside discourse. The 'claims to truth' that are advanced by science, religion and other discourses are considered by post-structuralists to be enforced by particular power relations.

Post-structuralism is also associated with the development of particular methodologies to explore these concerns. Derrida, for example, promoted the method of the deconstruction of 'texts' (that need not necessarily be written texts) as a means of destabilising truth claims (Norris 1982), while Foucault traced the genealogies of discourses to uncover their contingency (see Foucault 1966, 1969, 1979). These approaches have been adopted by a number of political geographers, notably in the field of critical geopolitics, while other political geographers have been attracted to the ideas of difference and of power and space that are prominent in much post-structuralist writing (see, for example, Deleuze and Guattari 1988; Foucault 1979, 1980, 1984).

Key readings: For an overview of the work of key post-structuralist writers see Lechte (1994). For a concise introduction to post-structuralist thought see Belsey (2002).

BOX 1.4 DISCOURSE

There are many different definitions of precisely what 'discourse' is, and the term is often used quite loosely in geographical literature. Put simply, however, discourses structure the way we see things. They are collections of ideas, beliefs and understandings that inform the way in which we act. Often we are influenced by particular discourses promoted through the media, through education, or through what we call 'common sense'. Derek Gregory, writing in *The Dictionary of Human Geography* (2000), identifies three important aspects of discourse.

1 Discourses are not independent, abstract, ideas but are materially embedded in everyday life. They inform what we do and are reproduced through our actions.
2 Discourses produce our 'taken for granted' world. They naturalise a particular view of the world and position ourselves and others in it.
3 Discourses always produce partial, situated, knowledge, reflecting our own circumstances. They are characterised by relations of power and knowledge and are always open to contestation and negotiation.

Key readings: Barnes and Duncan (1992) and Gregory (2000a).

arrangement with all the cells facing a central observation tower. The circle meant that prisoners could not see or communicate with each other, but also by means of backlighting from a small external window it allowed prisoners to be constantly visible via a large internal window from the observation tower, whose own windows had blinds to prevent prisoners seeing in. The prisoners could not know whether they were being watched at any particular time, but had to presume that they were under constant surveillance and therefore act within the rules. As Foucault describes,

> the major effect of the Panopticon [is] to induce in the inmate a state of conscious and permanent visibility that assures the automatic functioning of power. So to arrange things that the surveillance is permanent in its effects, even if it is discontinuous in its action; that the perfection of power should tend to render its actual exercise unnecessary; that this architectural apparatus should be a machine for creating and sustaining a power relation independent of the person who exercises it.
>
> (Foucault 1979: 201)

Although Bentham's panopticon was never actually built, the principle of control through visible yet unverifiable surveillance, assisted by spatial ordering, has been replicated in many areas of social and political activity. More broadly, the ideas about space and power that Foucault explored through his study of the panopticon have been translated into political geography through work on the ordering and control of space, for example, by Herbert (1996, 1997) on policing strategies in Los Angeles and by Ogborn (1992) on the exercise of state power in nineteenth-century England.

The influence of ideas from post-structuralist and postcolonial writers meant that the 'cultural turn' not only identified new avenues of geographical enquiry, but also introduced new conceptual and methodological approaches, including the use of discourse analysis to 'deconstruct' the meaning of texts, maps, policy documents and landscapes. However, as with Marxist political economy two decades earlier, the uptake of these innovations in established political geography was patchy. It was more commonly cultural geographers who took up the challenge of the new research questions posed by the cultural turn than people who described themselves as 'political geographers'.

Surprisingly, perhaps, the area of political geography where the new conceptual and methodological approaches had most impact was the neglected field of geopolitics. Drawing on Foucault's notions about discourse, as well as on critical political theory, geographers, including most notably Simon Dalby and Gearóid O'Tuathail, began to develop the new approach of *critical geopolitics*. By treating geopolitical knowledge as a discourse, critical geopolitics has sought to question, deconstruct and challenge geopolitical assumptions. This has involved, for example, examining the use of geographical metaphors such as 'heartland' and 'containment' in framing strategies, and, significantly, exploring the popular geopolitical knowledges that are constructed through cultural media such as film, literature, news reports and cartoons. We discuss critical geopolitics in more detail in Chapter 3.

The second recent influence on political geography has come from the development of feminist geography and from feminist theory more broadly. To date, few attempts have been made to think through an explicitly 'feminist political geography' (see England 2003; Hyndman 2001; Kofman and Peake 1990), but, engagements with feminism have highlighted the masculinist nature of traditional political geography and have begun to suggest ways in which political geography might be done differently. The conventional concerns of political geography have tended to focus on institutions such as the state, government and political parties which are dominated by men and tend to reproduce a male view of the political world (Drake and Horton 1983). Less attention has been paid to the institutions through which the patriarchal power of gender relations is exercised (such as the family) or to the spaces in which women's political activity has conventionally been focused – local education, health and childcare systems, the household and the voluntary sector. The integration of feminist perspectives into political geography has been associated with the development of work on the politics of 'public' and 'private' space, and on place/space tensions (England 2003; Taylor 1994a, 2000). England (2003: 611) proposes 'a feminist political geography that takes formulations of the politics of "public" and "private", power, space, and scale seriously', which she illustrates through a discussion of the political significance of scale for foreign domestic workers in Canada. Notably, the empirical research that England cites was not initially

designed as a political geography project (England and Stiell 1997; Stiell and England 1997), yet, as she suggests, there is much political geography that is implicit in previous work by feminist geographers.

Moreover, feminist theory and activism in general have challenged traditional notions of the 'political' that underpinned many essentialist definitions of political geography by proclaiming that 'the personal is political'. Combined with the influence of poststructuralism and cultural studies, this message has helped to change perceptions about the scope of political geography, extending the boundaries of the field far beyond those envisaged by many traditional definitions that focus on the state, or territory, or the analysis of political regions.

The future of political geography

Political geography is clearly a much more expansive creature today than it was twenty or thirty years ago. However, the danger of this transformation is that 'political geography' may become devalued by its very ubiquity – if everything is 'political' then it could follow that all geography is 'political geography'. This logic was followed by Clarke and Doel (1994), who employed post-structuralist theory and a Derridian writing style to imagine a 'transpolitical geography' which spilled over the limits of political geography's normal concerns and interests. The disturbing consequences of this proposal are posed in the accompanying commentary by Chris Philo:

> does this mean that swathes of work on the geographies of empires, states, nations, territories and boundaries (from Mackinder's geopolitics to Taylor's 'world systems') now become solely of historical interest, given that such work operates with the objects specified in a passing domain of politics? And does it also mean that much conventional research on administrative, electoral and locational conflict geographies might have to be waved goodbye as well? Clarke and Doel appear to answer in the affirmative.
>
> (Philo 1994: 529)

At the same time, the status of political geography has been challenged in more grounded terms by the fact that much of what might be considered as political geographical research is not being undertaken by 'political geographers' (Cox 2003; Flint 2003). Research on cultural politics and geography is performed by cultural geographers; on citizenship and the geographies of policy delivery by social geographers; on governance, regulation and state theory by economic and urban geographers – as well as sociologists and political scientists; on state formation and national identity by historical geographers; and on geopolitics by students of international relations.

These concerns have informed a debate about the future direction of political geography as a sub-discipline which was articulated in a panel discussion at the conference of the Association of American Geographers in Los Angeles in 2002 and a themed issue of the journal *Political Geography* in 2003. The perceived problem was expressed by Flint (2003), who pointed to the 'paradox' that while political geography (at least in the United States) was in good institutional health, it appeared to lack coherence and face uncertainty about its direction. Flint identified the uncertainty with the dilemma of whether political geography should concentrate on politics with a big 'P' or a little 'p':

> Identity politics, the environment, post-colonialism, and feminist perspectives are all relatively 'new' politics, placed on the agenda by the political upheavals of the 1960s . . . and can be classified as politics with a small 'p'. They stand in contrast to the old politics of the state and its geopolitical relations, statemanship or politics with a large 'P'.
>
> (Flint 2003: 618)

Flint argued that knowledge of both Politics and politics is required to understand the contemporary world, and that coherence could be maintained for 'political geography' by focusing on 'the way that different spatial structures are the product of politics and the terrain that mediates those actions' (p. 619), and by showing the relevance of spatiality to all types of power. Yet he also noted that much work on the

'new', small 'p' political geography is undertaken by individuals who are not 'card-carrying' political geographers, thus raising concerns about disciplinary boundaries that were echoed by Cox (2003). Other participants in the debate saw less cause for alarm. John Agnew, for example, emphasised the historic fluidity of political geography and commended its diversity with a geographical analogy:

> Much of what is of interest to me in contemporary political geography is exciting precisely because there is more limited agreement than was once the case in political geography and is the case today in some other fields (such as economic geography). By analogy, political geography is like Canada or Italy, a complex entity in imminent danger of collapsing under the weight of its internal differences. But for this very reason each is more interesting to the political geographer than, say, Luxembourg.
>
> (Agnew 2003: 603)

Broadly speaking, the debate produced three possible pathways for the future. The first is *concentration*, in which political geography would refocus on traditional key concepts such as the state (Low 2003) or territory (Cox 2003), reverting to an essentialist definition of the sub-discipline and establishing firm boundaries that distinguish it from cultural geography, economic geography and other predatory neighbours. The second is *expansion*, celebrating the dynamism and diversity of political geographical research and proactively seeking new objects of study as part of a 'post-disciplinary political geography' (Painter 2003). Kofman (2003), for example, argued that 'there isn't necessarily a contradiction between a heightened interest in political questions in human geography and the existence of something called political geography' (p. 621), while Marston (2003) noted that 'the migration of the political to other areas of the discipline seem to me to be compelling evidence that we have failed to attend to a large portion of what is legitimately and centrally the purview of political geography' (p. 635). The third pathway is *engagement*, forging new intellectual connections with allied subjects such as peace and conflict studies (Flint 2003), socio-legal studies (Kofman 2003), political ecology (Robbins 2003), feminist geography (England 2003) and political theory (Painter 2003), as well as with political geographies produced from outside the insular environment of Anglo-American geography (Mamadouh 2003; Robinson 2003).

We have already indicated that we are sympathetic to definitions of political geography that emphasise diversity, and hence to the pathways of expansion and engagement. This is reflected in the breadth of topics covered in this book. However, the key point to note here is the continuing dynamism of political geography. What we present is a snapshot of political geography at a particular moment in time, and even by the time you read these words new research will have been published, new debates started, new ideas proposed and new areas of study emerging. It is in this sense that this book presents an introduction to political geography, providing a foundation from which the student of political geography can engage with the cutting edge of the sub-discipline through journals and research monographs.

The structure of the book

This book is organised into three parts, each of which starts from a different perspective. Part 1, 'State, territory and regulation', starts with the state, which as we have noted above has conventionally been considered a key focus of political geography. The first chapter in Part 1, 'States and territories', examines the development of the territorial state and the significance of territory to the operation of the modern state. The next chapter, 'The state in global perspective', discusses the external relations of the state and the part that geography plays in them, including geopolitics. By drawing on a regulation approach to political economy, the final chapter in Part 1, 'The state's changing forms and functions', focuses on the forms and functions of the contemporary state and the strategies adopted by the state in the regulation of economy and society

Part 2, 'Politics, power and place', starts with place, a core geographical concept. The first chapter, 'The political geographies of the nation', considers the concept of a 'nation' and the ways in which national identity is linked with specific places and territories. The second chapter, 'Politics, power and place', steps down a scale to think about place as locality. It explores how place is important to politics and discusses the structuring of power within place-based communities. The final chapter in Part 2, 'Contesting place', examines how places become sites of political conflict, including conflicts about the meaning of symbolic landscapes and the construction of community.

Part 3, 'People, policy and geography', starts with people, but does so from two different directions. The first chapter in Part 3, 'Democracy, participation and citizenship', examines the ways in which people engage with the political process as citizens and how this engagement both is shaped by geographical factors and creates new geographies. The second chapter, 'Public policy and political geography', focuses on policy, the means by which the state engages with people in a place. This chapter discusses debates about the extent to which human geographers should engage directly with the policy process and raises issues that political geography students could consider in their own work.

A book such as this cannot hope to give any more than a flavour of the rich variety of topics that form part of contemporary political geography. As an introductory text, it is hoped, the book will stimulate you to read further on themes that interest you, and even to become involved in producing your own 'political geography' through undergraduate and postgraduate project work.

Further reading

Agnew's *Making Political Geography* (2002a) provides a more detailed history of political geography than that outlined here, albeit one which emphasises the traditional concerns of the sub-discipline more than recent innovations.

The debate about the future direction of political geography, discussed towards the end of this chapter, is published in *Political Geography*, 22, 6 (2003).

Many of the classic texts in political geography can still be found in university libraries, but it is often more informative to read more contemporary commentaries on these books and articles rather than the originals themselves. For more on Ratzel's theories, Bassin's paper 'Imperialism and the nation state in Friedrich Ratzel's political geography' in *Progress in Human Geography* 11 (1987), 473–95, is a good overview. O'Tuathail's paper 'Putting Mackinder in his place' in *Political Geography* 11 (1992), 100–18, is a similarly good source on Halford Mackinder. Herb, *Under the Map of Germany* (1997) is an interesting exploration of the perversion of cartography by German *Geopolitik*.

Kasperson and Minghi's edited collection *The Structure of Political Geography* (1969) contains reprints of many significant contributions from Aristotle onwards. Although it is long out of print, many university libraries will have copies. Agnew's reader *Political Geography* (1997) has an illustrative sample of more recent writing from the 1970s onwards.

Many of the themes explored by political geographers since the 1970s will be covered in more detail in later chapters and guidance to further reading will be given then.

STATE, TERRITORY AND REGULATION

States and territories

Introduction

Rather like the air we breathe, states are organisations that surround us as individuals, influencing and, in many ways, offering sustenance to the lives we lead. Similar to the air we breathe, states are also organisations that often lie beyond the limits of our critical reflection. We may question the priorities of political parties; we may also disagree with the policies implemented by various governments. We do not often question, however, the character of the organisation that political parties, while in government, seek to govern. In other words, we rarely think about what states actually are, how they are constituted, how they come into being and how they change over time. These are some of the questions concerning the form of the state that we will ask, and ultimately seek to answer, in this chapter.

The first and most fundamental issue we need to deal with, of course, is what exactly is a state? Fortunately, a number of eminent social scientists have sought to answer this question. Max Weber, for instance, argued that a state is a 'human community that (successfully) claims the monopoly of the legitimate use of physical force within a given territory' (Gerth and Mills 1970: 78). Michael Mann (1984) has built on this definition by arguing that any definition of states should incorporate a number of different elements:

1 a set of institutions and their related personnel;
2 a degree of centrality, with political decisions emanating from this centre point;
3 a defined boundary that demarcates the territorial limits of the state;

4 a monopoly of coercive power and law-making ability.

An extended definition such as this makes us think about a number of important aspects of the state. Significantly, it encourages us to think about the state in a far more abstract sense. In addition to the various paraphernalia associated with states in individual countries, there are, or at least there should be, certain underlying constants in states throughout the world, ones that are highlighted in the above definition. Rather than viewing the state in purely personal terms, therefore – as a supplier of public utilities, or something that is embodied in a senate or parliament, for instance – it makes us think of the underlying processes and institutions that (usually) help to constitute state bureaucracies.

So what do geographers, and more specifically political geographers, have to offer in any study of the character of states? In other words, what can we gain from studying the state from a geographical perspective? We argue that there are three main reasons for doing so. First, geographers can help to illuminate the fact that states, when considered at a global scale, vary from region to region. One geographer whose work can be said to demonstrate this is Peter Taylor (see Taylor and Flint 2000). Taylor has deployed the ideas of Immanuel Wallerstein (1974, 1979, 1980, 1989) regarding the existence of a capitalist world economy – one in which the northern states of the First World thrive through their exploitation of southern states and people of the Third World – in order to structure his understanding of the economic disparities that exist from one region of the world to another (see Chapter 1). Crucially, this process of exploitation leads to significant differences in the political, economic and social viability of Southern states on the 'periphery'. To put it bluntly, they do not have the money to support either their state institutions or their citizens. To states such as these, the badges of statehood – described by Mann – are ones to be aspired to, rather than being ones that reflect political reality. This is a theme that we will return to later in the chapter.

Second, geographers can help to highlight the unequal effect of particular policies on different areas within the territory of the state. These may range from policies that are explicitly targeted towards certain areas – for instance, policies of urban renewal in Europe and North America – to more general or 'national' policies. Even though these latter policies may be directed towards the state's territory and population as a whole, they may in turn have different impacts in different areas of the state, owing to pre-existing cultural, social, economic and political geographies. So, for instance, a policy to encourage the use of public transport amongst the general public through the raising of taxes on private car ownership may work successfully in urban areas, where opportunities for the use of public transport proliferate, but may further disadvantage the inhabitants of rural areas, where levels of car dependence are far higher. Obviously, geographers have a key role here in studying, and attempting to alleviate, these problems. This is a theme that we will discuss in Chapter 8.

Third, and most fundamentally, geographers can contribute to our understanding of the state because of the state's effort to govern a demarcated territory. Indeed, this is the main justification for studying the state from a geographical perspective and it is a point worth emphasising. Broadly speaking, states use the notion of territoriality in two main ways. In the first place, territoriality is important in a material or physical sense (see Brenner *et al.* 2003). This is a relatively straightforward idea and relates to the fact that states always try to demarcate the physical limits of their power. A good illustration of the growing power of the state over the modern period (since approximately 1500) has been its efforts to demarcate

its physical boundaries in a more precise manner. In this way, diffuse and ill-defined frontiers became precisely delineated boundaries (Newman and Paasi 1998). One of the best examples of this physical territoriality lies in the context of the shifting boundary between the US federal state and Mexico (see Prescott 1965: 77–87; Donnan and Wilson 1999: 34–9). For much of the nineteenth century a conflict existed between the two countries regarding their territorial extent. Much of the wrangling revolved around the precise location of the boundary between the two states. Critical here were natural features used to designate the boundary, such as the Rio Grande river. A large amount of the conflict concerned the changing course of the river over time and the implications this had for the territorial extent of the two states. A more recent set of examples revolve around the geographical limits of the Chinese Republic. Hong Kong, for a century an integral part of the British Empire, was reinstated into the Chinese Republic in 1997, and there have been long-term conflicts regarding the independent status of China's other independent neighbour, Taiwan (see Box 2.1). Obviously, these examples clearly illustrate the major importance of territory and boundaries in a physical sense.

In the second place, a state's territory is of key significance in an ideological context. What we mean by this is that the notion of territoriality is used by the state as a way of explaining its way of governing and ruling its population. In a sense, it is possible to argue that states do not govern people as such. Rather, they govern a defined territory and it is by doing so that they subsequently govern the people living within it. As Robert Sack (1983, 1986) has argued, this territorial method of control is in many ways a far easier way of governing than one that emphasises the direct control of people. Within this system, anyone living or working in, or even passing through, a state's territory is subject to the laws and policies of that state, regardless of their social, ethnic or cultural background. In this context, therefore, a state's internal and external boundaries are not only lines drawn on a map, ditches dug or stones laid on a barren moor. They are this, but they are also far more: they represent the ideological basis of state power.

BOX 2.1 STATE AND TERRITORY: CHINA AND TAIWAN

The uneasy relationship between China and Taiwan can be traced back to the period immediately following the Second World War. In 1949 the Communist Party of China defeated the Nationalist forces of Jiang Kaishek and gained control of the whole of the country, apart from the remote area of Tibet and the islands of Taiwan and Hong Kong. Since that period, the Chinese state has tried to gain control of these enclaves. Chinese subjugation of Tibet began in 1950 and Hong Kong was ceded from the British Empire to China in 1997. This has left Taiwan as the only remaining blot on the territorial integrity of China and the main focus of the Chinese state's enmity. Indeed, Taiwan has been a thorn in China's side for many reasons. First, Taiwan was the place of refuge for the Nationalist forces of Jiang Kaishek after the revolution of 1949. Second, Taiwan's status as an independent state has consistently been supported by the United States. The main reason for US interest in Taiwan was the perceived need to maintain a series of territorial footholds throughout the Asian continent. US troops were stationed on the island until 1978 and contributed to the frosty relationship that existed between the United States and China. With the US withdrawal from Taiwan, China has began to reaffirm its belief that Taiwan is an island that it can legitimately lay claim to. Much military posturing between the two countries – including missile testing and a series of military manoeuvres by the Chinese armed forces – has helped to reinforce the territorial significance of Taiwan. In this example, the state's key role in taking and maintaining physical control over a defined territory is clearly illustrated.

Key readings: Calvocoressi (1991) and Shambaugh (1995).

Our aim in this chapter is to explore the changing nature of the state over time. Adopting this historical approach enables us to demonstrate the development of some of the key features of state power. In order to accomplish this, we will focus first on the process of state 'consolidation' (Tilly 1990) that occurred from approximately the sixteenth century onwards. This gradual process led to the formation of the all-powerful states of the twentieth century. One useful way of exploring these changes is to examine the changing territoriality of the state, something we address later on. Here we also briefly discuss the arguments that suggest that states are being systematically undermined by the processes of globalisation operating in the contemporary world. The final theme we discuss is the process of 'exporting' the state from Northern to Southern states. This process has not been unproblematic, and we discuss the issues facing Southern states in detail. Taken together, the themes discussed in the chapter make us appreciate the importance of territory to the state, as well as the need to explore the temporal and spatial variations in state forms.

The consolidation of the state

In this section we discuss the development of the state over the past 400 years. We do not argue in this context that the state has existed only for this relatively short period of time. This is patently not the case. Ancient states existed as far back as 3,000 BC in Mesopotamia, the region occupied by modern-day Iraq (Mann 1986). We focus our attention here on the state from approximately 1500 onwards, since it is this relatively recent process of state formation that directly informs the political geography of contemporary states.

Charles Tilly (1975, 1990) has argued that fundamental changes affected the state from approximately 1500 onwards, most clearly in Europe. During this so-called modern period a series of far-reaching developments occurred in the nature of states as they gradually became 'consolidated' into their present forms. By 'consolidation' Tilly means the way in which the states of this period – mainly in Europe – became territorially defined, centralised and possessing a monopoly of coercive power within their boundaries.

Of course, this process of consolidation, which echoes some of the themes raised in Mann's definition of states, speaks of an earlier period when European states did not adhere to this organisational formula. The state of the earlier medieval period was a haphazard affair and included a plethora of different individuals and organisations claiming power over territory and space. These included kings, lay and ecclesiastical lords, religious organisations and free townspeople (see Anderson 1996). The process of consolidation that affected state forms during the modern period entailed the gradual abandonment of this medieval legacy and the development of states that possessed clearly defined territories, and which were capable of exploiting their land, people and resources in an efficient manner.

We can put some empirical meat on these bare bones by discussing specific studies that have examined the consolidation of state power. Miles Ogborn (1998), for instance, has explored some of the key changes to have affected the English and Welsh state during the seventeenth century. Crucial to the consolidation of state power during this period was its ability to collect revenue from its people, resources and land. The main method employed by the leaders of the English and Welsh state was the excise duties – or in other words, the taxes raised – on the consumption of beer. Furthermore, a series of new mechanisms were employed in order to ensure the efficiency and consistency of this process. At one level, this meant that a number of different officers were paid to survey the process of brewing beer. This involved much travel throughout the country to ensure consistency in the way in which beer was produced and sold. At a more fundamental level, efforts to raise excise duties on beer production led to sustained attempts to comprehend the internal geometry of barrels and casks. According to Ogborn (1998), by 'mapping' the internal 'geographies' of barrels, the state could ensure that excise duties were raised in an efficient and consistent manner. Efficiency was important, of course, since it enabled the state to raise as much revenue as it possibly could. The consistency and equality of collecting excise duties was just as significant, since it helped the state to legitimise the whole process to its citizens.

This example helps to draw our attention to an important feature of the consolidation of state power. States during the modern period were not solely concerned with collecting monetary resources from their people and territory, though this factor was without doubt crucial. Ogborn's study of the development of excise duties in early modern England and Wales is significant for another reason, for it emphasises the key role played by the collection of information and surveillance in the changing power of the state. In this case, the state had to know exactly how much beer was being produced. This general point has been well made by Anthony Giddens (1985). He has explored the crucial role of surveillance in the consolidation of state power. It is the act of collecting information, of recording it within the state bureaucracy, and of using it in order to govern a population, that is so critical to the consolidation of state power. Giddens (1985: 179) argues succinctly: 'as good a single index as any of the movement from the absolutist state to the nation-state is the initiation of the systematic collection of "official statistics".'

Key here is the notion of the infrastructural power of the state, or in other words, the power of the state to affect the life of its citizens in a routine manner (Mann 1984). The promotion of surveillance is both the product of, and the precursor to, the growth of the infrastructural power of the state. States require information about their population, their resources and their land in order to develop higher levels of bureaucratic control. In the same vein, the formation of state bureaucracies enables the development of more sophisticated means of monitoring the state's population and territory. This type of state–society relation is portrayed chillingly in George Orwell's famous novel, *Nineteen Eighty-four* (1949).

As well as collecting information about the state's population and territory, another important aspect of the growth of state power during this period relates to the production of certain knowledges. Michel Foucault (1977, 1979) has consistently argued that a key feature of the growth of state power during the modern period lay in the state's ability to produce knowledges concerning its population. So, for instance, before the state could impose more restrictive rules and regulations on the practices of its population, it first had to develop a series of knowledges concerning the difference between acceptable and unacceptable behaviour. In this way, considerable efforts were made to define deviance, various forms of sanity and insanity and of morality and immorality. Only by developing these sets of very specific knowledges could the state ensure that it was able to classify its population in the correct manner. In developing these knowledges, and in classifying its population, the state was then able to take action against it, for instance by building prisons, madhouses, workhouses, where criminals, the insane and the immoral could be taken (see pp. 12–13 above). In focusing on these issues, Foucault seeks to draw our attention to notions of 'governmentality' or, in other words, the rationality involved in government (see Box. 2.2). By this, Foucault means the development of a 'way or system of thinking about the nature of the practice of government (who can govern; what governing is; what or who is governed), capable of making some form of that activity thinkable and practicable both to its practitioners and to those upon whom it was practised' (Gordon 1991: 1). This process of developing knowledges and rational forms of government, therefore, further fuelled the growth of the infrastructural power of the state.

One useful way in which we can think of these changes to the nature of state power is through reference to James Scott's (1998) ideas regarding states' efforts to make the society and territory that they govern more 'legible'. States during the modern period faced significant problems in their efforts to govern in an effective manner. At least part of these problems lay in the complexity of the societies and territories that states sought to govern. For instance, Scott (1998: 27–9) notes that methods of measuring varied greatly throughout Europe during the early modern period (approximately between 1600 and 1800). Since there were so many different means of measuring, Scott argues there was considerable potential for geographical variation in the amount of taxes and excise paid by various communities, so much so that we can refer to a politics of measurement:

Even when the unit of measurement – say the bushel – was apparently agreed upon by all, the fun had just begun. Virtually everywhere in early modern Europe were endless micropolitics about how baskets might be adjusted through wear, bulging, tricks of weaving, moisture, the thickness of the rim, and so on . . . How the grain was to be poured

BOX 2.2 GOVERNMENTALITY

Governmentality is associated with the work of Michel Foucault and is essentially concerned with the problem of how government renders society governable (Foucault 1991; Rose 1993). 'Classical governmentality' hence involves both the description and problematisation of society, and the putting in place of techniques and mechanisms to respond to the problems identified. However, Foucault also proposed a second, more historically specific, meaning of governmentality which 'marks the emergence of a distinctly new form of thinking about and exercising power in certain societies' (Dean 1999: 19; see also Foucault 1991). This latter approach places an emphasis on the political rationalities that inform government, including the identification of the proper spheres of action of different types of authority. Once the legitimate scope of government has been established, the governmentality perspective also highlights the technologies of government, such as censuses, statistical surveys, maps and legal processes, and the apparatus of security, such as health, education and social welfare systems, that enables state power to be exercised.

Key readings: Dean (1999), Foucault (1991) and Rose (1993).

(from shoulder height, which packed it somewhat, or from waist height?), how damp it could be, whether the container could be shaken down, and, finally, if and how it was to be leveled off when full were subjects of long and bitter controversy.

(Scott 1998: 28; see also Kula 1986)

Key to the consolidation of power, according to Scott, were states' efforts to make the society and territory that they governed more 'legible'. What Scott means by this is the state's attempt to simplify the society that it governed and to make it more 'understandable' to the state's institutions. Instead of the great variety in methods of measurement, therefore, the state sought to impose its own standardised means of measuring. Perhaps the best example of this process were the efforts made by the French state during the eighteenth and nineteenth centuries to replace the various local and regional means of measuring distance with a rational and standard system of measurement based on the metre and kilometre. Box 2.3 discusses another instance of states attempting to make society and territory more legible in the context of the growth of scientific forestry in Prussia (modern-day Germany). In all these projects, linked with attempts to standardise methods of naming, measuring, growing and even brewing, we see the state's determined efforts to create a more legible, and therefore more governable and exploitable, population and territory.

Here, therefore, we see the efforts of the modern state to consolidate its power over the long term. This is the process that led to the shift from the so-called puny states of the medieval period, ones which were 'marginal to the lives of most Europeans', into the all-powerful organisations that are 'of decisive importance in structuring the world we live in today' (Mann 1984: 209). The main question that arises is this context is what motivated these massive changes in the nature

---

BOX 2.3 MAKING NATURE LEGIBLE: SCIENTIFIC FORESTRY IN NINETEENTH-CENTURY PRUSSIA

A recurring theme with regard to the consolidation of state power has been the attempts made by the state to exploit its territory and people in a more efficient manner. One particularly critical resource for the state during the modern period was its forests. These enabled it to support indigenous industry and also acted as the material for the construction of buildings and ships. Forests, though, were originally a problematic resource in that they were very disorderly in the way in which they were organised. In mixed forests, in particular, valuable timber was interspersed with trash and other varieties of trees and shrubs. For Scott (1998) this resource was especially 'illegible' in nature. Starting in Prussia (northern Germany) during the nineteenth century, however, a concerted effort was made to make this resource more legible, and therefore more manageable, for state foresters. New sampling and mapping techniques were developed that enabled foresters to quantify the amount of useful timber within a given area of woodland. Building on this, state foresters began to impose their own rationality on the forests of Prussia, so that they could create new forests, ones which were easier to 'count, manipulate, measure and assess' (Scott 1998: 15). Here, therefore, we see the beginnings of the regimented and uniform stands of trees, by today so familiar in many states. What is crucial here, however, is that this attempt to impose a certain rationality on a complex natural resource was linked with the efforts of states to consolidate their power. One key way of achieving this aim was to create a more rational, simplified, standardised and, as Scott puts it, more legible society and territory for the state to govern.

Key reading: Scott (1998).

of the state? What was the driving force behind the consolidation of state power?

Michael Mann has convincingly argued that the main reason for the consolidation of state power during the modern period was the need to conduct wars. Using empirical evidence from the English state, Mann (1988) has mapped out the changing patterns of the state's finances over the very long term. There has been a general tendency towards ever-increasing economic exploitation of the people and land of the English state. Critically, the main periods of growth correspond almost exactly with wars that the English state was involved in. Even during the limited periods of stability and peace that followed each conflict, state expenditure remained at a higher level than it had been during the pre-war period. This so-called 'ratchet effect' relates to the need to pay back the loans that were taken out by the state in order to sustain its war effort. Mann's work demonstrates clearly that the main impetus for the growing consolidation of power over the long term was the need to promote the state's ability to wage war against other states, and the equally crucial need to defend itself against aggressive neighbours.

An important theme, in this respect, is the close relationship between the external or international relations of states and their internal or domestic political geographies. This point was made over fifty years ago by one of the fathers of political geography, Richard Hartshorne (1950), but it is an idea that is also apparent in Ogborn's work (see Chapter 3). Ogborn has argued that it was the English state's attention to the domestic political and economic geographies of excise duty collection that enabled it to build external colonies that extended througout much of the known world. The same relationship between the external and internal political geographies of the state can be seen in the context of the interaction between the state's need to conduct wars and its efforts to tax its citizens in a more effective way (see Figure 2.1). The need for tax revenue had two main consequences. At one level, resistance to taxation could lead to increased surveillance of the population and the creation of a more repressive political regime. One could argue, for instance, that it is this resistance to war-induced

taxation that – at root – explains the formation of the police forces that exist in all states, as well as the development of ever more detailed systems of law and order. Paradoxically, the need to sustain a war effort could lead to the political and civic emancipation of the citizens of a particular state. Painter (1995: 45) has argued that the state's need to raise taxes in order to support its armed forces in many ways explains the creation of more representative democracies within the state. In other words, state leaders were forced by different factions within the state to pay a price for the financial support they received during times of war. As Tilly (1990: 64, original emphasis) has succinctly put it with regard to the English state of this period, 'Kings of England did not *want* a Parliament to assume ever-greater power; they conceded to barons, and then to clergy, gentry, and bourgeois, in the course of persuading them to raise the money for warfare'.

Warfare *between* states could, therefore, have both repressive and emancipatory consequences *within* each individual state (see Figure 2.1). The exact nature of these consequences depended in large part on the balance of power that existed within each state. (See Rokkan 1980 for a discussion of these varying alliances and conflicts within European states.) Moreover these consequences were socio-economically and geographically uneven in nature, depending on the ability of each social, economic, political and cultural faction to resist state repression and, furthermore, to mobilise the state machinery for its own ends. (In a general context see Jessop 1990a.) We discuss the relationship between the internal and external political geographies of the state in greater detail in Chapter 3.

The other key point that comes from Mann's work on the financing of the English and Welsh state over the long term relates to the function of the state as an organisation (see Clark and Dear 1984). Contrary to the work of some theorists, who view the state as an organisation solely concerned with furthering the process of capitalist accumulation, Mann's work demonstrates that states – for a considerable period in their history – have been concerned with fostering and sustaining the means to wage war on their competitors. Admittedly, states broadened their range of functions to deal with the regulation of capital accumulation and

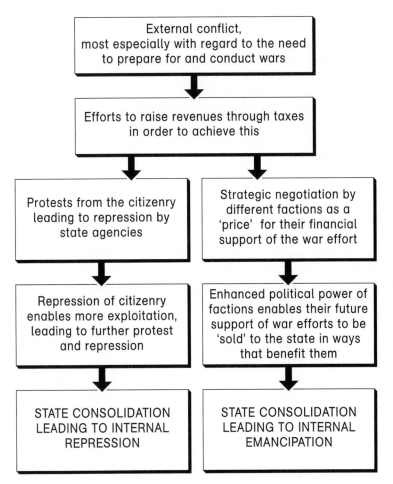

Figure 2.1 External and internal political geographies of the state

Source: after Tilly (1975); Mann (1988); Painter (1995)

related welfare issues during the nineteenth and twentieth centuries. This, however, is a theme that we discuss at greater length in Chapter 4.

In this section, we have discussed the consolidation of state power and have explored different theoretical contributions that have sought to explain this change. In the following section we explore some of these themes in greater detail by focusing explicitly on the growing territorialisation of the state. In addition, we will explore those processes — most particularly revolving around the forces of globalisation — that are allegedly undermining the territorial integrity of the state.

Building and challenging state territoriality

One of the main consequences — and driving forces — of the consolidation of the state during the modern period was the increasing emphasis placed by state rulers on governing defined territories. As discussed in the introduction to this chapter, territoriality plays a key role in the material and ideological constitution of every modern state. In other words, states have to govern territory in order to secure their physical form. Similarly states derive much of their legitimacy from the fact that they govern territory and not people. As Robert Sack (1986) has argued, the territorialisation of power can often cloud the repressive and exploitative nature of power relationships. This is part of the reason for the political success of states as they seek to govern groups of people.

A focus on the territoriality of the state helps us to answer a number of questions regarding state power. First, it enables us to think about one common aspect of the physicality of state forms, namely the extent to which most states during the modern period were of a medium size. Here it is useful to turn to the work of Hendrik Spruyt (1994), who has examined the changing territoriality of state forms in Europe during the modern period. Spruyt's key argument is that the European state has tended to gravitate towards a medium size and scale. Importantly, this was not, in any way, predestined to happen. Indeed, three different territorial formats were open to state rulers on the eve of the modern period: the city-state, such as Florence and Genoa; the extensive empires such as the Holy Roman Empire and the Hanseatic League; and the medium-sized state, such as England and France. Crucially, according to Spruyt (see also Tilly 1990), the medium-sized state possessed a distinct advantage compared with the other two possible territorial formats, for the reason that it offered the most appropriate combination of economic and military power. Both were needed in order to sustain successful war efforts. Generally speaking, cities are sites of the production and consumption of capital. They are, therefore, well suited to producing the economic resources needed to sustain a war effort. The economic power of artisans and guilds, however, made it difficult to coerce a large proportion of the inhabitants of cities into the state's armies. As such, city-states were not able to cope with the specific functions required of a state during the modern period. Large empires of the early modern period, however, were characterised by considerable coercive capabilities. They were filled with a warrior

nobility at the head of large populations. Unfortunately, these extensive empires did not possess the economic might to support the large-scale military activity needed of a modern state. In effect, Spruyt argues, a successful warring state would require a combination of these two factors – cities in order to produce capital and a warrior class leading large armies. Medium-sized states possessed this balance between cities, as sites of capitalist production, and large agricultural hinterlands, acting as the territorial basis for its warrior leaders.

Spruyt's work helps to draw our attention, therefore, to the key role played by a particular form of territoriality in promoting and sustaining the power of given states. States were far more likely to be successful in wars, and were far more likely therefore to survive as coherent political units, if they were of a particular size and scale. Of necessity, this point demonstrates the key role of geography in the constitution of states during the modern period.

Another key point made in the introductory section of this chapter regarding the territoriality of the state was the crucial role it played in sustaining the ideological integrity of the state as an organisation. Importantly, this territoriality has not been static in any sense. Alexander Murphy (1996) has explored these themes at length and he has suggested that we should distinguish between two interrelated aspects of state territoriality (see Figure 2.2). First, we need to think about territoriality as something that governs the relations between states: this, in other words, is the degree to which states within an international system of states live by a series of rules and regulations, responsibilities and obligations. This aspect of territoriality largely relates to the international or external activities of states such as in the context of war and diplomacy. Second, state territoriality also exists at a far more fundamental level in the way it relates to the 'relationship between territory and power in a sovereign state system; its central focus is the degree to which the map of individual states is also a map of effective authority' (Murphy 1996: 87). Murphy refers here to the internal territoriality of a state, or in other words, the extent to which a state possesses a practical and effective control over all its territory. As can be noted from the

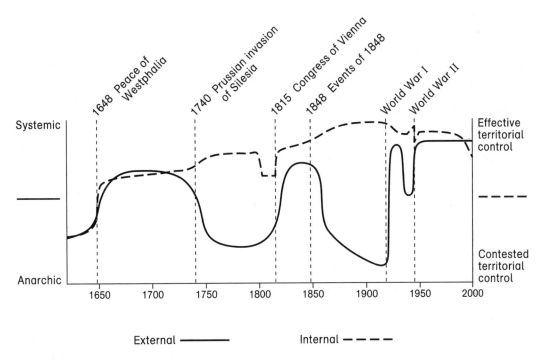

Figure 2.2 Changing 'international' and 'national' state territorialities

Source: adapted from Murphy (1996: figs 1–2)

schematic representation in Figure 2.2, there have been variations in the nature of state territoriality over time. Territoriality in an external or international sense has fluctuated wildly over time, ranging from predominantly stable or 'systemic' periods, when states, on the whole, respected each other's territorial integrity – for instance during the period after the Peace of Westphalia in 1648, or after the Treaty of Versailles in 1919 – to more unstable or 'anarchic' periods, or in other words, times of war and territorial encroachment. We can think here of the international anarchy characteristic of the First and Second World Wars.

Focusing our attention on the internal territoriality of states, the diagram portrays a far more stable pattern. Generally speaking, states have increased their ability to promote a more territorialised form of power over time. This relates back to Mann's (1984) ideas concerning the growing infrastructural power of the state over time. One of the main ways that states have ensured

that they can reach out and govern their citizens in an effective manner during the modern period has been through proclaiming that all people living, and all the land lying, within the boundaries of the state are subject to the state's laws and coercive powers. In effect, the majority of European states for much of the modern period have sought to create a homogeneous and isomorphic state territory. The aim for many states has been to reach a situation in which distance from the centre of the state has little bearing on states' ability to govern and rule.

The aim of creating a homogeneous state territory was a difficult one, for the simple reason that it meant changing age-old traditions and customs in the various localities of the state. It was often easier to achieve this goal in new lands, less structured by community traditions. It is no surprise, therefore, that some of the more successful attempts to impose a rational territoriality on state space have been achieved beyond the

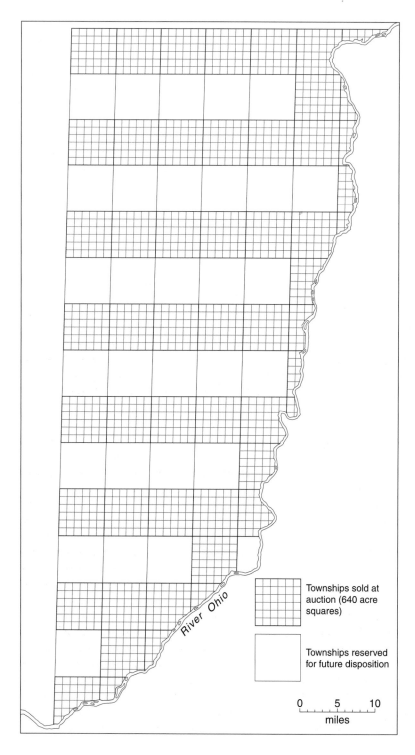

Townships sold at auction (640 acre squares)

Townships reserved for future disposition

0   5   10
miles

Figure 2.3 The imposition of rational territoriality in North America

Source: adapted from National Geographic (1988)

River Ohio

boundaries of Europe. In Figure 2.3, showing an early map of North America, we see a clear effort to create what J.C. Scott (1998) has referred to as a 'legible' society and territory. The map shows the precise partitioning of lands that took place to the north-west of the Ohio river in the late eighteenth century as territory was subdivided into townships and smaller parcels of land.

The efforts to create a territorial rationality for state space often involved a degree of conflict. In this regard, various options were open to state leaders who wanted to enforce their territorial order on state space. First, states could try to sponsor a process of centralisation whereby the political, economic, social and cultural norms of the centre were imposed on the state's periphery. Smith (1989) has argued that this was a significant feature of socialist states during much of the twentieth century. Socialist states' position as the prime directors of economic, political, social and cultural processes within their boundaries left little room for other actors – for instance, the forces of capital – to distort its vision of a wholly uniform state territoriality. For instance, the Soviet Union's monopoly of power enabled it to 'shape the spatial structure of society' in 'accordance with its own political preferences', based on notions of communism (Smith 1989: 323). This had enormous implications for the notion of territoriality within the Soviet Union, leading to sustained attempts to dissolve 'town–country' and 'region–region' differences within the country. In this 'grand plan' for the development of the Soviet state, little attention was paid to spatial variations in society and culture. The key factor that underpinned this drive was the vision of a homogeneous, and therefore 'legible', Soviet communist territory.

Second, the creation of a homogeneous state territory could occur through negotiation between the centre and the periphery. Ogborn (1992), for instance, has argued that many of the Acts adopted in Britain during the nineteenth century as a means of creating a more homogeneous British territory were developed through a process of negotiation between Parliament in London and the various localities. Acts such as the Police Bill of 1856 and the development of poor relief were fine-tuned as a result of political bargaining between centre

and periphery. In no sense here was a homogeneous British territory created through the uncontested imposition of the norms of the core of the British state on its periphery. Building on this, we can think of a third mechanism for a state, eager to govern its territory in a more effective manner, namely that of federalism. In a federal system, some powers are delegated to various regions within the state whereas others are maintained at a federal scale. As such, it can be said that federal states represent a somewhat emasculated form of territoriality: though some political and economic rights and responsibilities exist at a territorialised federal scale – thus forming a relatively homogeneous state space – others vary from province to province, creating a mosaic form of territoriality (see Box 2.4).

One final way in which territory helps to give shape to the power of the state is through the promotion of state nationalism, which engenders a sense of loyalty amongst citizens towards the state and its institutions. This is a theme we will discuss at much greater length in Chapter 5. It is important to note at this juncture, however, that nationalism, as an ideology, emphasises the link between a group of people and a certain territory. As such, territory can be viewed as the concept that unites the bureaucratic and impersonal organisation of the state and the political and cultural community of the nation (see Taylor and Flint 2000). It is another important reason for exploring the state – and the nation – from a geographical perspective.

We have discussed the key importance of territoriality for states during the modern period in this section. In the contemporary or late modern period, however, much is being made of the fundamental organisational changes that are affecting the state's functions. These changes are largely attributed to the forces of globalisation. As political, economic, cultural and social processes gravitate towards a broader, global scale the independence and autonomy of individual states throughout the world are allegedly being undermined. Importantly, these processes possess a territorial dimension. What we mean by this is that globalisation is often portrayed as something which is undermining the territorial integrity of states. If the states of the modern period can be described as 'power

BOX 2.4 TERRITORIALITY AND FEDERALISM IN THE UNITED STATES

One of the best examples of a federal state, incorporating a limited territorial homogeneity at a national scale, is the United States. European expansion into North America between the sixteenth and nineteenth centuries progressively signalled a radical re-evaluation of the meaning of space compared with indigenous peoples' understandings of the land in which they lived. Mirroring Scott's ideas regarding the state's need to make society legible, many charters were granted to Europeans that described their newly acquired land in a precise territorial manner. As Sack (1986: 136) notes, the charter granting land to William Penn in 1681 helped to define his land in a delineated manner. He was granted the land 'bounded on the East by the Delaware River, from twelve miles distance Northwards of New Castle Town unto the three and fortieth degree of Northern Latitude . . .'. At one level, therefore, there were considerable efforts to fashion a territorial basis for the new state that was forming in the United States. At a grander political scale, however, there were many debates regarding the exact way in which this new state should be governed. The main sticking point was the potential size of the new state, and the belief in the difficulty of governing a large state in a democratic manner. After independence in 1776, federalists, such as Madison and Jefferson, saw a federal state structure as offering the most democratic form of governance. According to these influential individuals, the smaller (provincial) states were favoured as the primary sites of governance, with the federal state scale limited to sectors that could not be governed at the scale of the individual state. In practice, this meant that the federal state's responsibilities were limited to defence and foreign affairs. As Jefferson argued (Sack 1986: 148), 'it is not by consolidation, or concentration, of powers, but by their distribution, that good government is effected . . . were we directed from Washington when to sow, and when to reap, we should soon want bread'. The federal structure adopted in the United States from this period onwards has, of course, profound implications for the nature of territoriality within the federal state. Federalism in the United States, owing to the devolution of considerable powers to various (provincial) states, has led to a limited territorialisation of power at the federal state scale, and an uneven pattern of territorial governance throughout the federal state.

Key reading: Sack (1986).

containers' (Giddens 1985), then there is some scope to argue that these containers are pitted with holes and are leaking badly (see Taylor 1995). We discuss the impact of globalisation on the state at greater length in Chapter 3. We think it important, however, to briefly address one aspect of the potential impact of globalisation on the territoriality of the state in this chapter.

The discussion in this chapter illustrates one important reason for countering the discourses that view the state as an organisation that is of irrelevance within the contemporary world, or one that has lost its territorial focus. These discourses often portray a simple shift from a state that was wholly territorial and homogeneous in nature to one which is increasingly

being challenged by global forces. What these interpretations conveniently forget is that states have never been constituted as homogeneous political entities. Even though the goal of individual states has been to promote a method of governance that is centred around a territorial ideal, they have often failed in this enterprise. Reasons for this failure are not difficult to countenance. One possible explanation is the lack of infrastructural co-ordination on the part of the central state to exactly map social practices within its boundaries. Linked with this are the attempts by the citizens of different states to contest the efforts of the state to regulate their behaviour in a territorial manner (see also Chapters 4 and 8). Even in states that seek to promote a wholly centralised notion of political power, there are numerous instances of the ability of citizens to challenge that power. France offers one example of this process. In his now famous book, *Peasants into Frenchmen*, Eugene Weber (1977) discusses the efforts of the French state to mould its citizens into one community of people. Though this is linked with the creation of a coherent French nation, we argue that it is also symbolic of an effort to forge a common and uniform territoriality for the French state. Importantly, however, this project was not altogether successful. Even during the nineteenth and twentieth centuries, people living in the various localities in France were able to challenge the centralising tendencies of the French state, for instance by preserving their own languages, dialects and customs. Another common way of challenging the territoriality of the state is through the act of smuggling, and this is a theme we discuss in Box 2.5.

This discussion shows that we need to question recent debates concerning the impact of globalisation on the territoriality of the state. As Yeung (1998) has demonstrated, states still play an active role in shaping the nature of political geographies within their boundaries. This means that state territoriality is still of crucial importance. In addition, state territoriality has never been unproblematic in nature. This fact

BOX 2.5 SMUGGLING AND STATE TERRITORIALITY: THE BORDER BETWEEN GHANA AND TOGO

The practice of smuggling is intimately bound up with state territoriality. By definition, smuggling depends on the organisation of the state: without state borders, smuggling would not have a reason to exist, since it is the act of illicitly transporting or trading goods, services or information across borders that constitutes the criminal offence of smuggling. In one way, therefore, smuggling helps to reinforce the importance of state boundaries. At another level, of course, smuggling demonstrates how porous state boundaries have always been. It is a practice that challenges the notion of a clearly defined and homogeneous state territoriality. We can see this process at work at the border between Ghana and Togo. During the colonial period, when Ghana and Togo formed part of the British and French empires respectively, attempts were made by the British to control the export of contraband from Ghana to its eastern neighbour. These attempts included the construction of customs posts along key routes between the two countries and the implementation of monetary fines for those caught with contraband. These efforts were at best half-hearted and there was a general acceptance that some illicit trade between the two countries was inevitable. Indeed, it was welcomed, since it brought a degree of financial and social stability to the border region. Under British control, therefore, it was accepted that the control of the state of Ghana was only partially constituted along territorial lines. With independence, considerable efforts were directed towards 'hardening' the border between the two countries. Far harsher measures were adopted in order to discourage smuggling, including introducing the death penalty for those caught smuggling gold, diamonds or timber. This situation did not last long, however, mainly owing to the ever increasing levels of smuggling between the two states and the evidence of the complicity of state officials. After a revolution in Ghana in 1982, the state realised that more lenient measures were needed to deal with smuggling, and it once again became a practice that was unofficially 'condoned'. In this example we see the uneasy relationship that exists between smuggling and the state: at one and the same time, they help to reinforce and challenge each other. As Donnan and Wilson (1999: 105) so aptly put it, 'smuggling . . . both recognises and marks the legal and territorial limits of the state and, at the same time, undermines its power'.

Key reading: Donnan and Wilson (1999).

makes it difficult to speak of globalisation as a process that is helping to undermine a purely homogeneous territory of the state, since it is unlikely that this uniform territorialised state has ever existed.

In this section we have discussed the importance of territoriality to the state. It is this aspect of the state that is the main reason for studying the state from a geographical perspective. In the final section of the chapter, we explore one other significant process that has happened over approximately the past fifty years. We refer here to the export of the state from its European, North American and socialist core area to the rest of the world.

Exporting the state to the world

The historical geography of the expansion of the state form to Southern countries is also an historical geography of the collapse of European empires. As the Spanish and Portuguese empires were broken up, largely during the late eighteenth century, and as the empires of Britain, France and the other European countries dissolved during the twentieth century, so were a large number of new, politically independent states formed in the three continents of South America, Africa and Asia. (See the newly independent states formed in Africa in 1960 in Figure 2.4.)

Primarily as a result of the imperial legacy, the leaders of these independent countries have sought to promote a political formation for their state that mirrors that of the various European metropolitan states. So, for instance, these new states have, on the whole, been organised territorially, and have incorporated a number of functions that are predominant in European and North American states (see Chapter 4). Vandergeest and Peluso (1995), for instance, have examined how state leaders in Thailand have sought to promote territoriality as a more effective way of controlling the country's people and resources. This project has unfolded in three main contexts: first, through the extension of a territorial form of civil administration into the rural areas of Thailand; second, through the promotion of survey-based land titles as the legitimate way to acquire and own land and, third,

through the demarcation of certain land as 'forests', which can then be acquired and exploited directly by the state.

Crucially, many Southern states have faced considerable problems in achieving these goals. We do not argue, here, that Northern states do not face significant challenges. We do argue, however, that the problems facing Southern states are of a different magnitude from those experienced in their Northern counterparts. Significantly, a number of the major issues facing Southern states revolve around notions of territoriality. Vandergeest and Peruso (1995) have shown the difficulties experience by the state in Thailand, both in promoting a uniform system of survey-based land titles and in defining large tracts of land as 'forest' to be exploited by the state. Major problems also arise with regard to Southern states' attempts to control their more peripheral land. Frontier regions within these states can often act as regional power bases for armed groups. We can think, for instance, of the *mujaheddin* rebels in Afghanistan, whose power base lay in the west of the country, the area farthest removed from the capital, Kabul. Other examples include the Tamil guerillas located in the northern Jaffna peninsula of Sri Lanka and the southern regions of India. Similarly, the portrayals that appear in the novels by Gabriel Garcia Marquez (1998) and Louis de Bernieres (1991, 1992, 1993) of the upper reaches of the Amazon basin, an area which lies beyond the immediate control and influence of the states of the region, are illustrative of the territorial challenges facing many Southern states. Here, life carries on regardless of minimal state interference, and there is a very strong sense of a 'peasant' life and community that lie beyond effective state control.

Linked with this theme is the extent to which Southern states possess a monopoly of coercive power within their boundaries (see the introduction to this chapter). This is patently not the case in many Southern states, which contain armed groups and even armies that curb the coercive power of the state. This process is most apparent in the case of civil wars. For instance, of the forty-five peacekeeping missions carried out by the United Nations between 1948 and 1997, only nine were located outside Southern states (Held *et al.* 1999:

Figure 2.4 The geographies of state formation in Africa: the miraculous year of 1960

Source: adapted from Arnold (1989)

126–9). This clearly illustrates the greater potential for military instability and civil war in Southern states compared with ones within the states of the North. This means, in effect, that Southern states cannot maintain their physical dominance over all their territory. For instance, after many years of surviving formal political opposition, the incumbent President of Sudan, Mohammad Nimeiry, who had presided over a gradual process of decay and decline within the country, was finally deposed only as a result of direct action in the form of a *coup d'état* in 1985 (see Calvocoressi 1991: 538). Once again, the notion of territoriality – so crucial in definitions of statehood – becomes problematic in Southern states.

In addition, Southern states often face severe problems in promoting a sense of civic nationalism – or in other words, a form of nationalism that is linked with the institutions of the state (see Chapter 5) – that can unite all its citizens. It is important to note at this stage that even the more stable and long-standing European states have experienced difficulty in sustaining their own civic nationalism. The growth of ethnic nationalisms in states such as Spain, France and the United Kingdom is testimony to this. This is an issue, however, that has far more potential for ethnic discord and violence within Southern states. One process that highlights some of these problems facing Southern states has been the substitution of many African nationalisms by ethnic or 'tribal' identities. The ethnic conflict and genocide in Rwanda between the Tutsi and Hutu ethnic groups is a disturbing instance of this tendency. In all, it is estimated that 800,000 Rwandans were massacred during the civil war in Rwanda, a war that had 'deep roots in politically fuelled inter-ethnic distrust and fear' (Wood 2001: 60). The formation of a Rwandan state with little cultural currency, along with the existence of a divisive political system, has hurried the descent into a political and ethnic abyss. What this example clearly demonstrates is the unstable foundation that exists for civic nationalism within Southern states.

These, then, are some of the problematic issues facing Southern states. Indeed, the extent and depth of these problems have led some political commentators, along with influential organisations, to argue that 'southern states . . . have no choice but to reduce their expectations of what, as modern states, they can do' (Hawthorn 1995: 141). At one level, such an argument is morally distasteful, since it opens the door to the potential neglect of the predicament facing Southern states by the states and organisations of the North. We disagree with this position on ethical grounds. At a more theoretical level, however, the quote raises interesting issues regarding the spatiality and temporality of state forms, and this is a theme we discuss in the conclusion to this chapter.

Timing and spacing the state

In this chapter we have stressed the changing nature of the state, over time and over space. States in Europe back at the beginning of the modern period were relatively puny organisations, with little ability to affect the lives of their citizens. With the process of state consolidation, this situation gradually changed as states became more sophisticated and powerful in their degree of infrastructural control. Importantly for us as geographers, we can link this shift with the growing territoriality of the state: governing a territory came to be viewed as the most efficient way of governing a given population. Contemporary globalisation, amongst other processes, is allegedly challenging the functions and the territorial integrity of the state (see Chapter 3). As far as we are concerned, the jury is still out regarding the effect of globalisation on the territoriality of the state. These debates, however, help to reinforce the notion that the state is not a static organisation, and is always undergoing changes with regard to its territoriality and, as we shall see in Chapter 4, with regard to its functions.

Similar arguments can be made concerning the spatial variation of state forms. This was true for much of the modern period and is especially true of the contemporary state. The main difference, especially since the end of the Cold War, has existed between Northern and Southern states. Partly as a result of their recently created status as independent states, and partly as a result of their lack of economic viability, many Southern states are struggling to maintain their territorial integrity. This again draws our attention to the changing nature of state form from one part of the world to another. A number of thorny questions arise with regard to this geographical patterning of state forms. Is the 'Northern' form of the state necessarily the model for 'Southern' societies? Is there a moral obligation on 'Northern' states and societies to interfere in repressive state forms that exist in 'Southern states'? Has the unstable state form of 'Southern' societies the potential to undermine the stability of states elsewhere in the world? These are not easy questions to answer, but they deserve our sustained consideration and action, not only as political geographers, but also as ethical and moral citizens of the world.

Further reading

The best starting point for an explanation of the significance of territoriality to the state is Michael Mann's paper 'The autonomous power of the state: its origins, mechanisms and results', *European Journal of Sociology*, 25 (1984), 185–213, reprinted in Agnew (ed.) *Political Geography: A Reader* (1997), pp. 58–81.

A good introduction to the changing nature of the state, especially in territorial terms, can be found in Anderson, 'The shifting stage of politics: new medieval and post-modern territorialities', *Environment and Planning D: Society and Space*, 14 (1996), 133–53. This paper discusses the changing character and importance of territoriality for the state over the long term and outlines the increasingly tangled territorialities of the state under globalisation.

An examination of the contemporary significance of territoriality to the state, and the associated impacts of globalisation on state territoriality, can be found in two interrelated papers by Peter Taylor: 'The state as container: territoriality in the modern world system', *Progress in Human Geography*, 18 (1994b), 151–62, and 'Beyond containers: internationality, interstateness, interterritoriality', *Progress in Human Geography*, 19 (1995), 1–15.

Fewer academics have attempted to chart the difficulties faced by southern states in promoting a territorial form of state power. An interesting study, however, can be found in Vandergeest and Peluso, 'Territorialization and state power in Thailand', *Theory and Society*, 24 (1995), 385–426.

The state in global perspective

Introduction

In the previous chapter, we discussed various aspects of the state's form. We focused on the historical development of the state and emphasised the significance of the state's efforts to control space as territory. In Chapter 2, therefore, we focused largely on the *internal* political geographies of the state. The aim of this chapter is to complement this discussion by focusing on the *external* relations of the state or, in other words, to examine the state in global perspective.

To some extent, a discussion of the state in global perspective may seem somewhat strange. After all, classical definitions of the state – as discussed in Chapter 2 – highlight the fact that states are organisations that seek to govern people, resources and land *within* defined boundaries. The portrayal of states that appears here is, therefore, relatively inward-looking and parochial. International and global issues would seem to matter little to the state, as defined by Weber and Marx. None the less, we began to illustrate in the previous chapter how all states look outwards beyond their defined boundaries in many ways. For instance, it was noted how much of the impetus to the consolidation of state power throughout the modern period came from the state's need to wage wars against its competitors. In this way, the state's interference in international matters could lead to far-reaching changes to its internal political geographies (see Mann 1988). This is not an isolated example of the close links between the internal and external geographies of the state. At a general level, for instance, it has been argued that many of the new means of governing that have been adopted in Northern states in recent years have

been 'borrowed' from other, apparently successful, states. In the age of 'fast policy transfer', states are on the constant look-out for new and innovative policies and practices to adopt within their own territory (Peck 2001). These two brief examples demonstrate clearly how states should be viewed as crucial actors on the international stage. In this way, we can see how the nation-state, existing at a *national* scale, also has *international* pretensions.

Statements such as these represent relatively well trodden intellectual ground within political geography. As noted in Chapter 1, Friedrich Ratzel, in many ways the father of the subject area, was keen to emphasise the need for states to actively engage in international relations. It was another key figure in the development of political geography, Richard Hartshorne (1950), however, who was to expound most systematically the need to focus on the external, as well as the internal, relations of the state. His classification of the state's external relations into the four different categories of territorial, economic, political and strategic is outlined in Table 3.1. Importantly, there is, according to Hartshorne, a close relationship between these four different categories, so much so that they should not be viewed as existing in isolation from each other. For instance, Hartshorne (1950: 127) notes how the interest of the United States in West Africa in the 1950s was shaped by a mixture of economic relations – significantly after the Firestone Tyre Company began rubber production there – and strategic – when West Africa was viewed as a convenient site for US airports during the Second World War.

Hartshorne's work is an important starting point for us in this chapter since it helps to illustrate the crucial

Table 3.1 The external relations of the state

Type of relation	Meaning	Example
Territorial	The degree to which neighbouring states agree on their respective boundaries. It also refers to the efforts made by states to regulate movement across their boundaries	The boundary between Finland and the former Soviet Union has long been disputed and speaks of uncertain territorial relations between the two countries. On the other hand, the movement of goods and people between the two states was highly regulated, especially during the Cold War
Economic	The economic trade that takes place between different states. This is important politically, since the economic resources that a state possesses affect the degree of its influence on international relations. Economic trade between different states can lead to the development of international alliances	The large deposits of the metal manganese, found in the Transcaucusus, in the former Soviet Union, was an important factor in the United States' relations with its Cold War enemy. Similarly, economic trade has acted as the stimulus to international alliances such as NAFTA and the European Union
Political	The degree to which a given state exerts political control or influence over other states. This can vary from direct political control, as in the form of a colony or empire, to more informal relations of influence and domination	The empire created by France in Africa, Asia, the Americas and Australasia during the modern period testifies to formal French domination of extensive lands throughout the world. Even though the majority of these lands are now independent, the French state still exerts informal influence on them
Strategic	Most fundamentally, the relationships that a given state chooses to enter into as a means of sustaining and enhancing its security or power. These relationships can change over time	The unification of the various German princiaplities in 1871 into one state encouraged the formation of a number of new strategic partnerships between other European states as they sought to counter the military threat posed by a united Germany

Source: after Hartshorne (1950)

significance of international political geographies for each state. A given state does not exist in isolation and therefore must engage with other states or institutions beyond its boundaries. It is, therefore, imperative for us as political geographers to examine the state in global perspective. We do so in two main contexts in this chapter. First, we discuss the way in which states – at certain times – can *extend* beyond their boundaries to engage with political, economic, cultural and environmental processes at an international or global scale. We focus here on two main themes: the link between states and the process of imperialism and colonialism; and the key role played by states within global geopolitical patterns. Second, we address the way in which certain global processes can *penetrate* the

boundaries of a given state. The substantive focus here is on globalisation and its impact on the state's territorial form. Although we concede that there are many commonalities between these three processes, we focus on them in different sections in this chapter for the simple reason that they have traditionally been discussed in separate spheres of geography. The reader will, none the less, appreciate the strong connections between the three topics.

States and empires

One of the most fundamental ways in which states have sought to extend their power and influence beyond their immediate boundaries has been in the context of conquering and maintaining control of overseas territories. This close relationship between states and empires was a feature of much of the modern period. As was discussed in Chapter 2, from about the sixteenth century onwards the nation-state began to assert itself as the most important political unit on the world map. At the same time, nation-states also began to attempt to control other lands throughout the world. This occurred in the context of the creation of empires.

The relationship between states and empires is, at first glance, relatively straightforward. Empires are said to be constituted as a result of formal and/or informal relationships of domination between different states. For instance, empires, especially in the form that they took from the nineteenth century onwards, have been described as 'an extensive group of states, whether formed by colonization or conquest, subject to the authority of a metropolitan or imperial state' (Jones 1996: 155). Similarly, the political relationship of imperialism, which is often based on the existence of empires, has been described as follows: 'The creation and maintenance of an unequal economic, cultural and territorial relationship, usually between states and often in the form of an empire, based on domination and subordination' (Clayton 2000: 375). These two definitions illustrate the conceptual relationship between states and empires. Empires are formed by a collection of states and so empires usually exist at a far grander spatial scale than states. In the same way, processes of state formation take place *within* particular states, whereas efforts to create empires are examples of inter-state activity and international relations.

States become the hub of extended empires through a mixture of political, economic and cultural/religious domination. Political relationships between the metropolitan state and its dependent territories can take on a variety of forms, ranging from a situation in which the empire is constituted as a unified political whole to one that is characterised by far looser forms of political co-ordination. Moreover, there has historically been very little rhyme or reason to the character of the political relationships adopted within particular empires. This was largely a result of the piecemeal manner in which empires were generally formed and

the impact that pre-existing political and economic relationships in the various conquered territories could have on later imperial developments. So, for instance, the British Empire of the late nineteenth century contained a mixture of Crown colonies, governed directly by the British Colonial Office, self-governing white-settled colonies, semi-independent protectorates and the Indian subcontinent, administered by its own India Office. As Morris (1968: 212) argues, the British Empire of the time was 'all bits and pieces. There was no system.' Similarly, the French Union during the period 1946–58 comprised metropolitan France, French overseas departments, territories and settlements; UN trusteeships; French colonies, which became overseas departments of France; and associate states or protectorates, such as Vietnam, Laos and Cambodia (Calvocoressi 1991: 478). Here, once again, the political geographies of empire were complex, containing a mishmash of relationships between the metropolitan state and its dependent territories.

The practical implementation of political control could range from the more mundane aspects of colonial administration to instances of physical coercion and oppression. Effective administrators, for instance, could improve the day-to-day running of imperial territories by seeking to regulate the social relationships that existed between settlers and the indigenous population. Political control could also be evidenced in the context of military coercion of indigenous populations, something that was facilitated by the advanced technologies of the European powers. There is no better testament of this than the battle of Omdurman in Sudan in 1898 when British and Egyptian forces confronted a Dervish army of 40,000 people and managed, with the aid of the Maxim machine gun, to kill approximately 11,000 of them (Abernethy 2000: 100). Although the idea of political control within the European empires of the modern period could mean many things, it was at its most tangible and unforgiving in the form of the muzzle of a gun. A more recent example is the use of, or threat of the use of, a mixture of 'smart' and cluster bombs by the US military in extending its imperial influence into the Middle East.

Economic relationships and trade also played a significant role in constituting the relationship between state and empire. Indeed, some have argued that it is these economic relationships between different parts of the world – often organised into empires – that has been the driving force behind much of world politics during the modern period. Key here are the ideas of the sociologist Immanuel Wallerstein (1974, 1980, 1989) regarding the existence of a capitalist world economy and Lenin's writings on the nature of imperialism (in Lenin 1996). As argued in Chapter 1, much of what sustained the imperial expansion of the modern period, according to Wallerstein, was the need of European states to discover new sources of cheap labour in different parts of the world. Some have been critical of Wallerstein's ideas, most specifically his attempt to explain world politics through sole reference to economic processes (see Giddens 1985). Lenin's work is also crucial, in this respect, and is central to economic, development and Marxist geographies of imperialism. Lenin conceived of imperialism as the final stage of capitalism, associated with the rise and crises of finance capital in the late nineteenth century. He argued that 'Imperialism emerged as the development and direct continuation of the fundamental attributes of capitalism in general' (Lenin 1996: 89). In this way, Lenin's work also emphasises the importance of economic relationships for the process of empire building.

Once again, economic relationships between the metropolitan state and its imperial territories could take on a variety of forms. Abernethy (2000: 57–63), for instance, has noted how economic relationships within empires could be organised either vertically or horizontally. Vertical economic relationships were part of the classic mercantile ideology, prevalent from the sixteenth century to the eighteenth, where trade occurred solely between the metropolitan state and its colonial territories. Raw materials, such as precious metals and spices, were extracted from imperial lands and served to increase the financial resources of the metropolitan state. Semi-finished products, such as luxury clothing materials and porcelain, also flowed from the Americas and Asia to the European metropolitan centres. Importantly, this was a two-way flow of trade, as European states exported raw materials and manufactured goods to their peripheral lands.

Abernethy (2000: 58), however, argues that the 'most significant trade patterns . . . violated the mercantilist ideal by being lateral, not vertical', as European states sought to link non-European lands in a global trading circuit. The most important of these horizontal patterns of trade, especially in the context of early empires, was the so-called 'slave trade triangle' (see R. King 1995: 10–15): ships from ports in Britain and France set sail for Africa with manufactured goods; these goods were traded with West African leaders for slaves; the slaves were then taken to the Americas to be sold; the money raised from this transaction was then used to buy raw materials and goods produced in the Americas, which were finally transported back to the metropolitan European states to complete the trading triangle (see Figure 3.1). It is international trade, such as this, linking a European core with an African, Asian and American periphery, that symbolises most clearly Wallerstein's (1974) notion of a capitalist world economy organised around the former's economic and political needs. Clearly, therefore, the economic geographies of trade were crucial to the development and sustenance of European empires.

European empires of the modern period were also predicated on cultural geographies of domination. Social and cultural differences are central to theoretical conceptions of colonialism and imperialism. Watts, for instance, defines colonialism as 'the establishment and maintenance of rule . . . by a sovereign power over a subordinate *and alien* people' (in Johnston *et al.* 2000: 93, our emphasis). Critically, ideas of cultural and social difference could help to justify the whole project of imperialism. In this respect, colonists could be led to believe that their imperial efforts were part of a process of 'civilising' an uncivilised 'other' (Said 1978, 1993) (see Box 3.1). Moreover, empires helped to reinforce ideas of difference. This is seen most clearly in the context of the emphasis placed on categories of race (Stoler 1991). Ideas of white superiority and discrete racial categories were reproduced in the context of empire as colonised peoples came to be viewed as second-class citizens (Jackson 1989). Morris (1968: 131–2) illustrates well the importance of ideas of race within the British Empire – and of social and cultural difference – in the following quotation:

The great ideal of Roman citizenship was only half-heartedly approached by the British. In theory every subject of the Queen, whatever his colour or skull formation, enjoyed equality of opportunity, and fifty years before Lord Palmerston, springing to the defence of Don Pacifico, a Greek merchant of Portuguese Jewish origin but British nationality, had almost plunged Europe into war. There was nothing to stop an African or an Indian going to Britain and becoming a bishop, a peer of the realm or Prime Minister. In practice, however, it was a racialist empire—what was Empire, Lord Rosebery had once rhetorically asked, but the predominance of *race*?

Ideas of cultural domination were clearly exhibited in the context of expansion of European state religions outwards into new imperial lands. The close connection between the religious activity of the Church and the political practices of states during the modern period must be emphasised. Protestantism and Catholicism during this period should be viewed, first and foremost, as *state* religions, intimately linked with the imperial project. For instance, the vast majority of religious

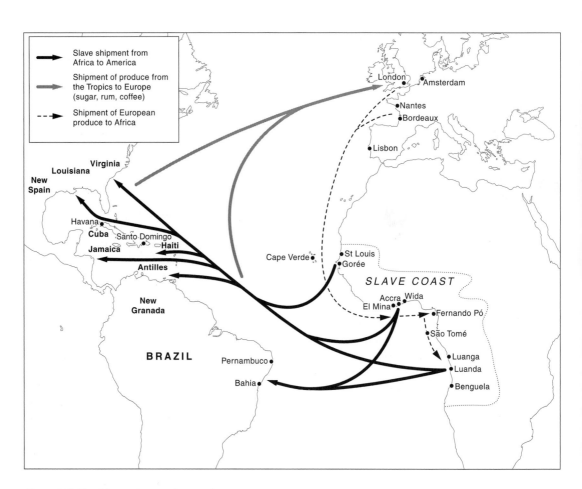

Figure 3.1 The Atlantic slave trade triangle
Source: adapted from King (1995: fig. 1.2)

BOX 3.1 EDWARD SAID AND THE NOTION OF 'OTHERING'

In his now famous book *Orientalism* (1978) Edward Said attempted to outline the relationship that exists between empires and ideas of cultural and social difference. Key to this whole process of domination is the idea of 'othering'. In describing the allegedly different qualities of peoples encountered on the periphery – which, in the majority of cases, involved emphasising their perceived weaknesses – Europeans could morally justify their efforts to civilise and exploit them. This process helped to essentalise racial and ethnic categories in both non-European and European lands. What this means is that particular racial or ethnic categories were perceived as homogeneous groupings of people, with internal differences being underplayed. There are many instances of this process. Black people, for example, could be described in a gentlemen's magazine of the late eighteenth century as follows: 'The Negro is possessed of passions not only strong but ungovernable; a mind dauntless, warlike and unmerciful; a temper extremely irrascible; a diposition indolent, selfish and deceitful; fond of joyous sociality, riotous mirth and extravagant show...a terrible husband, a harsh father and a precarious friend.' The significance of such a description, according to Said, is that the alleged negative characteristics of the black person were seen to represent the direct opposite of the white person's alleged strengths. For Said, it is this cultural act of domination that justified many of the worst excesses of European state imperialism of the modern period.

Key readings: Said (1978, 1993).

personnel within the Spanish empire of the sixteenth and seventeenth centuries were of Spanish birth. Moreover, they were appointed by the Spanish court so that their actions were controlled by the political arm of the state (Abernethy 2000: 231). It is not surprising, therefore, that the Protestant and Catholic religions of the various imperial European states were often used as a way of reinforcing ideas of cultural domination within the empire. Europeans, as members of either the Protestant or the Catholic Church, were in receipt of religious salvation and, therefore, could be easily distinguished from a 'savage' condemned to the fiery halls of hell. The relationship between state politics and state religion, however, was rarely a straightforward one. Religion could destabilise the rigid social and cultural boundaries erected between colonisers and the colonised. Once a member of the Christian faith, indigenous individuals in various colonised lands could be viewed by the Church as the equals of the European colonisers. Indeed, Christophers (1998) has shown this to be the case in nineteenth-century British Columbia, where a resident missionary, John Booth Good, viewed

coloniser and colonised alike as either sinful or saved individuals. In this way, religion at one and the same time helped to reinforce ideas of social and cultural difference and to destabilise them.

This section has discussed the political, economic and cultural relationships of domination that were so important to the European empires of the modern period. Although it has been useful to address these three sets of relationships separately, it should not surprise the reader that the different spheres of domination were often intimately related. This is clearly the case in the example discussed above, which illustrated the close linkages between the Spanish state and Catholicism within the Spanish empire. Similarly, the United States has used a mixture of military intimidation, economic might and cultural dominance as a way of facilitating its recent imperialist pretensions. The connections between different sectors of the French empire are discussed in Box 3.2.

This discussion has shown the different tactics used by European states of the modern period in order to further their political control of lands throughout the

BOX 3.2 SPHERES OF DOMINATION WITHIN THE FRENCH EMPIRE

The French empire of the modern period demonstrates clearly the way in which different forms of domination coalesced within the imperial project. In the scramble for land in the Americas, Africa, Asia and the Pacific state officials, soldiers, merchants and missionaries combined to extend French control. A cadre of state and military officials provided the political domination so crucial to the empire-building process. The merchant class was critical to the French empire's economic domination of its conquered lands and religious orders, such as the Pères Blancs of the nineteenth century, could provide a much needed cultural focus of domination. One pattern – evidenced in the expansion of French power from Senegal to the West African interior during the nineteenth century – was for state officials and the armed forces to exert physical control over new lands, and for business interests and the Church to advance once the indigenous population had been pacified. Another model saw the united advance of state officials, the armed forces, business interests and religious orders, but, as Abernethy (2000: 235) argues, 'with a clear understanding of which actors would undertake which tasks'. This method of empire building was evident in the French expansion into Canada. Whatever the method employed, the key point is the fact that co-operation between different sectors of the metropolitan state was crucial for a successful process of French empire-building.

Key reading: Abernethy (2000).

world. Political control, economic exploitation and the emphasis on social and cultural difference all combined to create a situation in which European states had managed to control, at some time, the vast majority of the Americas, Asia, Africa and Oceania.

Of course, this formal political control of non-European lands was not to last. Lands in the Americas gained their independence from the Spanish, Portuguese and British empires by the early nineteenth century. A second round of decolonisation took place in Australasia, Asia and Africa during the twentieth century (see Taylor and Flint 2000: 116–19). It has been argued that this has not led to the creation of completely independent non-European states. Indeed, it has been suggested that some dominant states still exert as much influence over non-European lands and peoples as they have ever done. In this world of informal influence and coercion, it is the geopolitical power exerted by states over other political actors that is important and it is this theme that we discuss in the following section.

States and geopolitics

As discussed in Chapter 1, there has traditionally been a strong association between geopolitics and political geography, though it has varied in strength over time (see Box 3.3). What is crucial for the present discussion is the way in which notions of geopolitics demonstrate another important, yet more informal, way in which states can reach beyond their boundaries to affect processes occurring at international and global scales. Our aim in this section is to examine the contours of this engagement, at both a conceptual and an empirical level.

Geopolitics is concerned with the manifold ways in which states seek to exert power and influence beyond as well as within their boundaries. The first question that needs to be addressed, therefore, revolves around the methods used to achieve this political domination. Here, it is useful to distinguish between the politico-geographical knowledges needed to 'make sense' of the world and the actions taken to sustain the political position of a given state within it. We will discuss each of these issues in turn.

The first element refers to conceptions of space, power and politics that inform the policies adopted by given states. At a general level, we need to stress the importance of ideas of realism within the international relations of various states (see Dodds 2000: 37–42). Realism within international relations is based on the assumption that all states are in active competition, one with another. As such, any given state should be distrustful of the activities of other states. Following on from this, the main aim of any given state is to maintain and develop its political status in the face of other states equally concerned with their own self-interest and status. It is in a political scenario such as this that the geopolitical considerations of states – in the form of military security and international political influence – become important. Despite the appellation of 'realism' for this body of theory, many commentators have argued that this is far from a 'real' interpretation of the character of world politics (see, for instance, Walker 1993). None the less, realist ideas have characterised states' attitudes towards international relations since the nineteenth century.

Within this broad realist framework, none the less, individual states must develop more specialist knowledges of other competing or allied states. It is here that the role of political geography becomes

---

BOX 3.3 THE RISE, THE FALL, THE RISE AND THE POSSIBLE FALL OF GEOPOLITICS

The notion of geopolitics has helped to shape the nature of political geography over time. The term came to prominence during the late nineteenth and early twentieth centuries and referred to the way in which ideas relating to politics and space could be used within national policy. The growing importance of the term during this period was not an historical accident. In the period subsequent to the 'scramble for Africa', there were few opportunities for additional European territorial expansion and, in such circumstances, international politics became increasingly focused on 'the struggle for relative efficiency, strategic position, and military power' (O'Tuathail 1996: 25). It was in this world that political geographers could aid state leaders in their efforts to increase the political influence exercised by individual states on the global stage. This period of geopolitical involvement in statecraft reached its apogee in Germany during the 1930s and 1940s, where ideas concerning the need for German territorial expansion were easily incorporated into Nazi ideology (Parker 1998: 1). Of necessity, perhaps, the period subsequent to the fall of that regime witnessed a waning of the star of geopolitics, both within the subject of political geography and, to a lesser extent, within policy circles. The re-emergence of geopolitics as a legitimate frame of enquiry took place during the 1970s, particularly in the United States and France (Parker 1998: 1). Its use during this period was very much based on the all-pervading, yet largely unconsummated, conflict between 'East' and 'West' that characterised the Cold War. Here again, it was the need for international political alliances, and the political geographies of influence that underpinned them, that acted as the much needed 'shot in the arm' for geopolitical debates. Geographers were to contribute to these. Since the mid-1980s, however, classical geopolitics has, once again, come under fire, in academic circles at least. Rather than supporting international and national political structures of domination, political geographers, affiliated to the subject area of *critical* geopolitics, are beginning to question and undermine these structures and the discourses and ideologies that surround them (see O'Tuathail 1996). Depending on one's perspective, therefore, this has either signalled another downturn in the fortunes of the notion of geopolitics within geography or has re-energised it in exciting and radical new ways.

Key readings: Dodds (2000) and O'Tuathail (1996).

important in helping states to 'visualise society'. The *locus classicus* of a political geographer shaping state policies is the contribution of Halford Mackinder, Professor of Geography at Oxford University at the beginning of the twentieth century. As we suggested in Chapter 1, his now famous 'heartland thesis' (Mackinder 1904), in particular, sought to advise the British state concerning the geographical balance of power within Europe and the wider world. Contained within this thesis were ideas concerning the shift from the importance of maritime to terrestrial bases of power and the growing threat from particular regions – Central Europe for instance – and states, especially Russia (see Figure 3.2). Despite the growing critical engagement with geopolitics within political geography in recent years (for instance Agnew and Corbridge 1995; Dodds 2000; O'Tuathail 1996), more conventional forms of geopolitical knowledges are still produced by key actors within political geography (see Box 3.4).

The application of these geopolitical knowledges and visions lies within the purview of state leaders. The strategic objectives of states can be achieved in many ways but we concentrate in this section, for the sake of brevity, on the overtly political, economic and cultural aspects of geopolitical engagement. In much the same way as the previous section, we would want to stress that these three are closely interrelated, even though we consider them separately here.

Political domination can take on many forms. At its most basic and uncompromising, it is based on military relationships between two or more parties. Much of the rationale behind the proliferation of nuclear weapons during the Cold War, for instance, was based upon the West and the East's need to secure strategic military and, therefore, political advantage over their enemies. This became the main justification for the global political and military face-off between East and West that characterised the international relations of the Cold War. A more recent example has

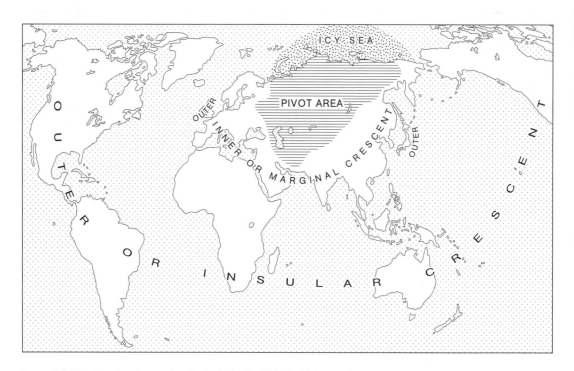

Figure 3.2 The 'heartland' map of Halford Mackinder (1904: 426)

BOX 3.4 GEOGRAPHY AT WEST POINT MILITARY ACADEMY, NEW YORK

West Point Military Academy, the institution that has trained army cadets in the United States for over 200 years, is probably the best remaining instance of the interaction between geography, geopolitics and statecraft. Here the emphasis of a geography career is clearly on addressing geopolitical and strategic issues. The departmental Web site, for instance, asserts that the discipline's focus is on 'the continued security of the [US] nation . . . We believe understanding and appreciating the world around you adds to quality of life Americans enjoy, which is an important reward of the freedom we serve to protect (http://www.dean.usma.edu/geo/gene.htm)'. In this respect, the publications of the Academy give a flavour of the way in which geography as a discipline is viewed. They include a CD-ROM entitled *Afghanistan: a Regional Geography* and another electronic publication, *Understanding International Environmental Security: a Strategic Military Perspective*. The former is significant in light of the recent United States-led military incursions into Afghanistan whereas the latter attempts to chart the possible impact of global environmental issues on the national security of the United States. These publications, along with the general ethos of the Department of Geography at the West Point Military Academy, illustrate a clear concern with geography's contribution to geopolitical and strategic issues. In this context, they represent some of the few remaining examples of the influence of conventional geopolitics on the subject area of political geography.

Key reading: The department's home page at http://www.dean.usma.edu/geo/gene.htm.

been the nuclear stand-off between India and Pakistan over the disputed province of Kashmir (Dodds 2000: 103–6). Once again, overt displays of the military might of the two countries have been used as a means of securing strategic, military and political advantage within the region. Political forms of geopolitical domination can also occur in more subtle and hidden ways. A good instance of this is the persistent military influence of the United States in neighbouring countries in the Caribbean, Central and South America (see Dodds 2000: 57). The most infamous examples of these more covert efforts by the United States to influence the internal politics of other independent states have been in Guatemala, Nicaragua and Cuba.

These latter examples also begin to demonstrate the strong connections between political and economic aspects of geopolitical strategy, where political interference is accompanied by various forms of financial aid. A key method of securing geopolitical influence and dominance in recent years has been the financial and technological aid offered by dominant countries to other, needy countries. In many ways, if military might represents the 'stick' of international relations, then financial aid is the 'carrot'. Numerous examples exist to demonstrate the role of economic influence in shaping international geopolitical relations. In the period after the Iraqi invasion of Kuwait in 1990, for instance, there was much debate in the international community concerning the best way to secure the freedom of the latter. Much of the political shenanigans of the period took place in the corridors of the United Nations in New York. The famous journalist John Pilger (1992) has noted how the United States tried to use its economic muscle as a way of securing the support of other states for its plan to mount an invasion of Kuwait and Iraq. In this respect, its main efforts were directed towards the non-permanent members of the Security Council of the United Nations, which, at that time, included one of the poorest states in the world, Yemen. It is a little-known fact that Yemen voted not to support an invasion of the Middle East by American-led UN forces. In the immediate aftermath of the vote, it is alleged by Pilger (1992), the Yemeni ambassador to the United Nations was informed by his US counterpart that that was the most costly decision he had ever made. In the following weeks, $70 million of proposed US aid to Yemen was cancelled, the World Bank and the International Monetary Fund began to question the economic practices of the Yemeni state and 800,000 Yemeni workers were expelled from Saudi Arabia.

As Dodds (2000) has argued, occurrences such as these are part of a broader range of economic strategies that help certain Northern states to achieve geopolitical dominance over Southern countries. The influence of industrialised countries over institutions such as the World Bank, the International Monetary Fund and the World Trade Organisation has been particularly important. It has helped to generate an additional layer of compliance within international relations. The best example of this process is the so-called 'structural adjustment programmes' of the World Bank, which seek to constrain the range of economic and political policies that can be pursued by less industrialised countries (Dodds 2000: 17; see also Krasner 2001: 28–9). The criticism levelled at these programmes is that they reify a particularly industrialised model of development on southern states and, as such, represent a new form of informal imperialism by northern states. In many ways, these examples illustrate the strong connections between geopolitics and the broader international political economy (see Agnew and Corbridge 1995).

The third element that helps to constitute geopolitical dominance is the cultural messages that shape our geographical and political understanding of the world. The field of critical geopolitics in recent years, in particular, has attempted to draw our attention to the value-laden messages contained in political speeches (e.g. Agnew 1998: 116) and more popular forms of culture (e.g. Sharp 1993). Indeed, much has been made of the intimate connections between the geopolitical imaginings forged in both formal and informal contexts. The most notable example of this interaction is the alleged influence of the war film *Rambo* on US foreign policy under the leadership of Ronald Reagan (Sharp 1999: 186).

The similarities between the role of culture within geographies of imperialism and geopolitics are strik-

ing, in this respect. In the same way as ideas of the essentialised categories of difference could be used to justify acts of colonial exploitation within formal empires, so can the propaganda contained within political and popular accounts of countries, cultures and religions help to create positions of geopolitical dominance. There is no better example of this process than the western reaction to the terrorist attacks that took place on 11 September 2001 (see Box 3.5).

As stressed earlier, political, economic and cultural aspects of international relations should not be viewed in isolation. Rather, they coalesce and reinforce one another to form the geopolitical patterns of power that are familiar to us today. Indeed, much has been written of the 'new world order' that has been created around the geopolitical might of the United States (see Williams 1993; see, however, Arrighi and Silver 2001). We need to realise, however, that the current geopolitical ordering of the world represents merely the latest manifestation of global patterns of power within international relations. Agnew and Corbridge (1995: 19–23), for instance, have usefully delineated three periods of geopolitical order within international relations (see Table 3.2), ones that have structured international politics for the past 175 years. Their work is important, in this respect, since it demonstrates the changing nature of geopolitical patterns over time. Although relationships of political, economic and cultural domination may continue for extended periods of time, they are by no means wholly stable and continuous. It is unsurprising, therefore, that many commentators argue that we have entered a fourth, unipolar period within international relations, subsequent to the final period noted by Agnew and Corbridge (1995), dominated by the geopolitical might of the United States (e.g. Ikenberry 2001).

States, therefore, seek to influence political, economic and cultural processes operating beyond their immediate boundaries. The one note of caution that needs to be sounded, however, is that some of the geopolitical literature can give the impression that states are wholly effective and single-minded in their pursuit of geopolitical goals. One fact that can offer some solace to individuals and organisations critical of states' seemingly self-serving interference in

BOX 3.5 GEOPOLITICAL IMAGES AFTER THE TERRORIST ATTACKS OF 11 SEPTEMBER 2001 ON NEW YORK

One of the clearest recent examples of the significance of cultural messages for global patterns of geopolitics came in the wake of the terrorist attacks on the United States that took place in September 2001. President George W. Bush, for instance, was keen to use images and rhetoric appropriated from the culture of the American west, referring to the need to 'smoke out' terrorists 'holed up' in the caves of Afghanistan. Famously, there was much disagreement about how to conceptualise the terrorist threat to the United States. By describing the United States as a civilised country of freedom and democracy, commentators in the United States were seen by many to be describing the states or peoples supporting terrorism as uncivilised. That the terrorists themselves can be considered uncivilised is not especially controversial but there was a too common assertion that all Islamic states should be viewed as uncivilised when compared with a civilised West. The most extreme example of statements like this came from Silvio Berlusconi, the Italian Prime Minister, who asserted that 'the West is bound to occidentalise and conquer new people', thus presumably leading to the dissolution of all Islamic states. Berlusconi's viewpoint was seen to be unhelpful for the formation of a coalition of states united against the threat of international terrorism, especially since the coalition would be strengthened immeasurably by the inclusion of moderate Islamic states. As a result, the United States was keen to portray Al-Qaeda as an organisation supported by one 'rogue' state, Afghanistan. Berlusconi, however, was not the only person to use unhelpful images and phrases during this period. 'Operation Ultimate Justice', the original title used to describe the US-led attack on the Taliban and Al-Qaeda in Afghanistan, was objected to by Islamic clerics on the grounds that ultimate justice can be dispensed only by Allah. This, once again, had the potential to antagonise Islamic members of the coalition against terrorism and, as a result, the offensive was renamed 'Operation Enduring Freedom'. These various examples demonstrate the key significance of cultural messages and images for forging geopolitical visions of the world.

Key readings: Harvey (2003) and Mann (2003).

international affairs is that the geopolitical goals of any given state can be, and often are, contradictory and self-defeating. This is due, at least in part, to the fact that any given state is not a unitary phenomenon, but is rather made up of a series of often competing institutions (Jessop 1990a). The upshot of this is that the geopolitical priorities of various sectors of any given state can contradict one another. A good instance of this geopolitical conflict of interest is discussed in Box 3.6.

Another more recent example of geopolitical contradictions is the tangled relationship between western states and Iraq. In supporting Iraq during its conflict with the neighbouring state of Iran, western states contributed to the military arsenal available to Saddam Hussein and his armies during their incursion into Kuwait. Of course, these weapons of mass destruction, partly supplied by western arms manufacturers and governments during the 1980s, are now causing much geopolitical consternation within the global coalition against terrorism. All in all, such an example demonstrates the potential pitfalls of seeking to impose a geopolitical order on an unstable and changing world.

This section has explored another way in which states can extend beyond their boundaries in order to shape political geographies at international and global scales. The following section – focusing on the relationship between globalisation and the state – looks at an apparently different process, whereby international and global processes penetrate the boundaries of the state.

Table 3.2 The changing patterns of geopolitical power, 1815–2002

Locus of power	Period	Forms of geopolitical dominance
Phase 1 Britain within a wider Europe	1815–75	Britain leads a European territorial expansion into non-European lands. Britain, partly because of its territorial expansion but also its sea power, holds a position of military dominance within the European continent. Britain uses this military dominance as a way of promoting particular global economic relationships, such as free trade and comparative advantage, that sustain its pre-eminent geopolitical position
Phase 2 The destabilisation of British power, especially as a result of the rise of Germany	1875–1945	National economies become more protectionist and there is far less emphasis on free trade. There is a struggle for geopolitical dominance, most clearly in the context of the two world wars of the twentieth century
Phase 3 The Cold War between East and West	1945–90	The world is divided into two spheres of geopolitical influence – East and West – based on the conflicting geopolitical might of the United States and the Soviet Union. The tentacles of these two countries extend well beyond the confines of Europe, the traditional battleground of geopolitical dominance, to incorporate all states in the world
Phase 4 A unipolar world, with the United States as its hub, based on ideas of free trade	1990–?	The Unites States, since the collapse of the communist bloc, tries to dominate world politics through a mixture of political and military influence, the economics of free trade and neoliberalism, and through the dissemination of numerous cultural messages. In many ways this dominance is similar to Britain's during *Phase 1*, though, as Agnew and Corbridge (1995: 20) argue, the United States is far more aggressive in its pursuit of global free trade and international political influence than Britain was, even in its heyday

Source: after Agnew and Corbridge (1995: 19–23)

BOX 3.6 THE CONTRADICTIONS OF GEOPOLITICS: BRITAIN AND ARGENTINA, 1945–61

A fine example of the potential that exists for contradictory tensions to develop within geopolitical relationships is the friction and conflict that existed between the United Kingdom and Argentina in the 1950s and early 1960s. There have been strong trading links between the United Kingdom and Argentina for some time – manufactured goods were exported from Britain and raw materials, like beef, were exported from Argentina to the United Kingdom. During the 1950s and early 1960s, however, tension rose between the two countries, mainly because of friction concerning the political status of the Falkland Islands and Antarctica. Despite these problems, the United Kingdom was keen to continue exporting goods to Argentina. A major conflict of interest arose, however, when the Argentine government wanted to acquire Shackleton aircraft from the United Kingdom. Certain sectors of the UK state, such as the Treasury, were of course, in favour, since it would have helped the UK balance of payments. The Foreign Office was ambivalent about the transaction. At one level, the sale of Shackleton aircraft would have helped to sustain the good trading relations that existed between the United Kingdom and Argentina. At another level, the sale could have weakened the political and military influence of the United Kingdom in the south Atlantic and for this reason the Ministry of Defence attempted to veto the sale. This example shows that because the state cannot be considered in the singular, so its geopolitical posturings cannot be viewed as singular either. This creates the potential for disruptions and contradictions within the geopolitical relationships espoused by any given state.

Key reading: Dodds (1994).

States and globalisation

As discussed briefly in Chapter 2, there has been much discussion concerning the alleged impact of globalisation on the state. Before examining these various debates it is important to define precisely what we mean by globalisation (see Box 3.7). Key here are the writings of Peter Dicken (1998, 2003) and Ash Amin and Nigel Thrift (1994, 1997).

There are, broadly defined, three different interpretations of the impact of globalisation on the state. The first interpretation views globalisation as something that is undermining the state and its territoriality, leading to new global forms of political organisation. For these so-called 'boosters', such as Kenichi Ohmae (1996: 5), 'traditional nation-states have become unnatural, even impossible business units in a global economy'. Proponents of this argument suggest that globalisation is associated with a 'borderless world' where states are relegated to a minor supporting role on the global stage of capital accumulation. Important political figures have supported this position. In the words of the British Prime Minister, Tony Blair, 'what is called globalisation is changing the nature of the nation-state as power becomes more diffuse and borders more porous' (*The Times*, 20 July 1995). For former US politician Robert Reich 'modern technologies have made it difficult for nations to control these flows . . . of knowledge and money' (1992: 111). This viewpoint has been popularly characterised by images

of McDonald's and Coca-Cola becoming part of the staple diet of people living in the vast majority of the countries of the world (see Plate 3.1).

Others have criticised the booster school, rejecting the argument that globalisation is a new phenomenon that is transforming contemporary political geographies. 'Hypercritics' (Dicken *et al.*, 1997) or 'sceptics' (Held *et al.*, 1999) – such as Hirst and Thompson (1996) – use historical statistics of world flows of trade, investment and labour in order to underplay the significance of more recent global patterns of trade. They claim that because levels of economic interdependence remain stable, no significant changes have occurred since the nineteenth century and it is, therefore, wrong to suggest that the contemporary period has witnessed the formation of a particularly integrated economy. For Hirst and Thompson, globalisation is a 'myth', used as a politically convenient rationale for practising neoliberal economic strategies. Furthermore, from this perspective, the nation-state remains a controller of cross-border activity and it is misleading to suggest that it has lost all meaning and significance within the contemporary world.

In between the 'booster' and 'sceptic' schools lies the middle ground occupied by numerous commentators (such as Dicken 1998, 2003; Held *et al.* 1999; Yeung 1998), who argue that globalisation should be viewed as a complex set of interrelated processes that are leading to a qualitative reorganisation of the geo-economy and nation-states. The main thrust of these arguments

BOX 3.7 GLOBALISATION

Globalisation, when considered in political, cultural and economic contexts, implies:

1 the increasing importance of financial markets on a global scale;
2 the centrality of knowledge as a factor of production;
3 the internationalisation and 'transnationalisation' of technology;
4 the rise of transnational corporations;
5 the intensification of cultural flows;
6 the rise of 'transnational economic diplomacy'.

(Amin and Thrift 1994, 1997)

When these various factors are considered together, something is certainly 'happening out there' in relation to the scale of political, economic and cultural activity (Dicken 1998). Crucially, we need to distinguish between 'internationalisation' and 'globalisation'. The former refers to the extension of political, economic and cultural activity across national boundaries. This has occurred since the formation of nation-states during the modern period (see Chapter 2). Globalisation, on the other hand, means 'not merely the geographical extension of [political, cultural and] economic activity across national boundaries but also the functional integration of such internationally dispersed activities' (Dicken 1998: 5). The key significance of globalisation, therefore, is the existence of behaviour, rules and organisations that normalise and regulate political, economic and cultural activity on a geographical scale greater than that of the nation-state.

Key readings: Amin and Thrift (1994, 1997) and Dicken (1998, 2003).

is that the nation-state is presented as 'permeable' to the processes of globalisation but, none the less, has a 'political complexion' that enables it to filter various global forces (Dicken 1998). In a similar vein, Henry Yeung (1998: 292) challenges the 'borderless world' discourse on the grounds that it plays down the 'intricate and multiple relationships between capital, the state and space'. There is, for Yeung, an enduring importance of national boundaries because capital is still embedded in distinct national social and/or institutional structures. According to authors such as these, the nation-state, therefore, remains engaged in mediating both domestic and transnational political, economic and cultural activity.

It is this third interpretation of the character of the relationship between the state and globalisation that we advocate in this book. States are being transformed in many ways as a result of the forces of globalisation. They still, however, play key roles in structuring

political, economic, cultural and social geographies within and beyond their boundaries. This point has been forcefully made by Yeung (1998). Focusing on the economic impacts of globalisation on contemporary states, Yeung has argued that states still retain instrumental roles in shaping the processes that take place within their boundaries (see Table 3.3). States, for instance, create the political, social and cultural conditions which may foster the global economic success of their home industries. The success of the Japanese electronics and car industries in the period between 1975 and 1990, for example, was in large part due to the correct political and economic institutions put in place by the Japanese state. Key, in this respect, was the way in which the Japanese state covered the costs of innovation within these industries in order to sustain their economic competitiveness. Equally important is the fact that some transnational corporations – seen by some as the antithesis of state

Plate 3.1 Booster accounts of globalisation: the spread of American foods

Courtesy of Gillian Jones

territorial power – are controlled directly by the state. The clearest instances of this phenomenon are to be seen in Europe. The French state, which controls international corporations such as the finance company Crédit Agricole, the petrochemical company Elf-Aquitaine and car manufacturer Renault, is a particularly good example.

A further weakness in many contributions to the debate on the nature of the relationship that exists between contemporary states and the processes of globalisation is that they tend to position states and global processes as two opposing combatants, involved in a 'zero-sum game'. What this means is that in the battle between states and the forces of globalisation there is an implicit assumption that each represents totally distinct and different categories. Furthermore, so the argument goes, if one becomes more powerful

and important then the other must automatically decrease in importance by the same amount. As Yeung's (1998) work demonstrates, this is clearly not the case. For instance, some key transnational corporations are owned and run by states. In addition, the United Nations – although a body that operates at a global scale – exists as a collection of individual nation-states. Another good example of the way in which nation-states can contribute to the forces of globalisation lies in the context of the World Trade Organisation (WTO). Even though this has been considered to be the antithesis of the nation-state, it is significant that it is an organisation formed by nation-states and run, ostensibly, for the benefit of nation-states (see Box 3.8).

Reports of the death of the nation-state are therefore exaggerated: states demonstrably still play a key role

Table 3.3 States and economic globalisation

The continuing role of states	Recent examples
States continue to act as the 'guarantor of the rights of capital': in other words, they ensure that capitalist accumulation is carried out according to certain rules and regulations	The US state during the 1980s actively lobbied Japan on behalf of its semiconductor industries in order to open up the Japanese market for US semiconductor transnational corporations
States create the domestic conditions for the success of their own transnational corporations	For much of recent history the Japanese state has covered the costs of new innovations for transnational corporations in certain sectors of the economy, especially automobiles and electronics
States are often directly involved in global economies through their ownership of certain transnational corporations	Examples include the French state's ownership of transnational corporations such as Elf-Aquitaine and Renault, and the many corporations that have strong links with the government of Singapore
States have a key role in regulating the world economy, so that they can influence the types of foreign investment that take place within their boundaries	China regulates the types of foreign firm that may invest within its boundaries. This involved confining them to Special Economic Zones for much of the 1980s
States play an important role in helping to set up international political and economic organisations that help to co-ordinate and regulate the global economy	States – and state politics – are instrumental in the decisions made by the World Trade Organisation, and other organisations such as the G7, the group of most industrialised countries
States still have an influential role in shaping their own domestic economies. These domestic economies still represent a large proportion of total economic activity in the world today	The implementation of particular policies within certain states may lead to changes in the nature of the domestic economy. Ronald Reagan's reduction of taxes as a means of increasing consumer demand within the United States can be seen as an example

Source: after Yeung (1998)

in structuring the processes that occur within their boundaries. Moreover, they also help to shape the character of those global processes that are allegedly undermining the power of the state.

Of course, this does not mean that states are not affected by the forces of globalisation. Indeed, there is strong evidence that they are being transformed in important ways. There are two key points to make in this respect. First, the impact of globalisation may be different in different states located in different parts of the world. Much has been written concerning the differential impact of the processes of globalisation in different parts of the world. Dodds (2000), for instance, has argued that for many Southern states the current

debates on the impact of globalisation on the state merely represent the recent concerns of Northern countries. For the vast majority of Southern countries, the political, economic and cultural dominance of transnational and global forces has been a feature of their national politics for tens, if not hundreds, of years. Following on from this, we can argue that the impact of certain aspects of globalisation varies from state to state, being dependent upon the underlying political, economic, cultural and environmental geographies of those states. In this regard, it has been noted that one particular aspect of globalisation – namely the use of the internet – has not impacted overmuch on Southern countries, because of the need for certain

BOX 3.8 GLOBALISATION, NATION-STATES AND THE WORLD TRADE ORGANISATION (WTO)

The WTO was formed in 1995 as a way of ensuring the expansion and regulation of free trade throughout the world. It is based on the earlier General Agreement on Tariffs and Trade, which also sought to regulate trade between states. The WTO has been commonly understood as an organisation that represents the antithesis of the democratically elected governments of nation-states. Viewed as a 'tool of the rich and powerful', the international messenger of transnational corporations and the opponent of progressive health, environmental and development policies, it has incurred the wrath of anti-globalisation activists. The most significant illustration of the hatred that has been directed towards the WTO was the often violent demonstrations that took place during the WTO summit in Seattle in 1999. One of the main criticisms of the WTO was that it reflected the imperatives of big business and, therefore, represented a wholly undemocratic organisation that had the power to influence not only international trade but also labour rights and degrees of environmental protection within particular countries. The irony, in this respect, is that the WTO is an organisation that is ratified by the governments of the states which are its members. At its heart, therefore, the WTO is not an organisation that exists over and above nation-states. Rather, it should be viewed as a global institution that is formed through the amalgamation of nation-states. This point does not diminish the environmental, political, social and economic injustices that have been committed as a result of WTO decisions over its brief history. Furthermore, the undue influence of certain powerful countries and organisations on WTO decisions is not to be welcomed. The important point we would stress, however, is that nation-states, in this context at least, are intimately involved in the production of new global political and economic institutions. As such, the processes of globalisation do not just happen 'out there' beyond the reach of the state. The state is actively involved in shaping global politics.

Key reading: de Burca and Scott (2001).

expensive technologies in order to surf the web (see P. Crang 1999). The fact that languages other than English are spoken in a large number of states also, of necessity, means that the impact of English as the lingua franca of globalisation varies from state to state.

Second, we want to emphasise that the loss or transformation of state power in one particular context need not mean that other aspects of state power are lost or transformed in the same way, if at all, even within the same state. Michael Mann (1997) has done much to emphasise this point. He has argued, for instance, that the integrity and territoriality of Northern states may well be weakened as a result of the processes of economic globalisation, particularly within the European Union. Here, the creation of the euro zone has challenged the power of many European states to make decisions concerning their own national interest

rates. On the other hand, a focus on the changing role of identity politics – the way in which people's politics are shaped by aspects of their identity, most notably their national identity (see Chapter 5) – demonstrates that Northern states are being strengthened, rather than weakened, under globalisation. It is this variable impact of different aspects of globalisation on the state that enables Mann (1997: 472) to argue that 'these patterns are too varied to permit us to argue simply that the nation-state and the nation-state system are strengthening or weakening' under globalisation. In effect, globalisation impacts on different states in different ways. Furthermore, these various impacts are context-specific.

The discussion in this section shows clearly that the impacts of globalisation on the state are complex. Moreover, it is not simply a situation in which forces at scales over and above that of the nation-state are undermining or transforming the political geographies of the state. As has been demonstrated, states are actively involved in the process whereby these global forces are forged in the first place. In this respect, rather than thinking of the relationship between global forces and the state as one that reflects a one-way and top-down process, it should be viewed more as a dialogue between the two. States, since their formation, have been actively involved in shaping the political geographies that exist beyond their borders. The relationship between states and current processes of globalisation demonstrates that this urge to influence extraterritorial political geographies seems set to continue.

Further reading

Three main themes are discussed in this chapter and the literature that addresses them is vast. For a good introduction to the link between states and empires, see Abernethy, *The Dynamics of Global Dominance* (2000). This book gives a good account of the various methods used to enhance state control of overseas land during the modern period as well as providing some interesting case studies of this process. A fascinating examination of the association between the discipline of geography and imperialism can be found in Godlewska and Smith, *Geography and Empire* (1994).

Many political geographers have contributed to our understanding of geopolitics and critical geopolitics. See, for instance, Agnew, *Geopolitics* (1998), Agnew and Corbridge, *Mastering Space* (1995) and Dodds, *Geopolitics in a Changing World* (2000). An account of the problems involved in geopolitical interference can be found in Dodds, 'Geopolitics in the Foreign Office: British representations of Argentina, 1945–61', *Transactions of the Institute of British Geographers*, 19 (1994), 273–90.

Economic geographers have tended to explore the impact of globalisation on the state, even though the various contexts within which globalisation impinges on the nation-state are numerous. Dicken, *Global Shift* (2003), offers a comprehensive and brilliant introduction to the theme of globalisation. For an account of the impact of globalisation on the state see Yeung, 'Capital, state and space: contesting the borderless world', *Transactions of the Institute of British Geographers*, 23 (1998), 291–309, and M. Jones and R. Jones, 'Nation states, ideological power and globalisation: can geographers catch the boat?', *Geoforum* (forthcoming).

The state's changing forms and functions

Introduction

The previous chapter concluded by discussing the ways in which the nation-state plays a key, although modified, role under the conditions and challenges of globalisation. This chapter continues on this theme by examining the changing institutional forms and functions of the capitalist state. It does this by drawing on a *régulation* approach to political economy and the state. The term 'political economy' is frequently used to discuss the interrelationships that exist between economic, social, and political processes, which are forged through power relations as 'moving parts' (see Peet and Thrift 1989). The *régulation* approach has a neo-Marxist take on political economy that stresses the ways in which capitalism is managed through state, economy and society 'interactions' (Florida and Jonas 1991). Box 4.1 introduces capitalism and summarises the differences between Marxism, structural Marxism and neo-Marxism approaches to political economy. This chapter, therefore, suggests that state institutional forms and functions can be explored in relation to the ways in which states are embedded or 'integrated' into different economic, social and political processes.

The political geographer David Reynolds, writing during the early 1990s, remarked upon the increasing number of geographers drawing on approaches such as *régulation* theory to understand 'the behaviour of states as economic and geopolitical actors at a variety of territorial scales' (Reynolds 1993: 389). This work was considered important because it was taking political geography into new territory and giving progress to its intellectual development. We discuss the state's changing forms and functions from this perspective

by first focusing on the origins of *régulation* theory and analysing its main arguments. Because the *régulation* approach is a challenging set of literatures, we then use case studies of how regulationist authors have applied this approach in their work and in doing so we draw out the changing institutional forms and functions of the state. The case studies analyse the *régulation* approach in relation to: economic and industrial geography; the geographies of scale within the context of neoliberalism; state intervention through modes of governance; the dynamics of urban politics and citizenship; and the purported shift from the welfare state to the workfare state.

A rough guide to the *régulation* approach

The *régulation* approach emerged from a particular strand of French thinking during the early 1970s. Researchers at the Centre for Mathematical Economic Forecasting Studies Applied to Planning (CEPREMAP) in Paris were faced with an interesting set of problems that could neither be resolved through conventional economic planning nor explained using existing theories of political economy (Jessop 1990b). Between the late 1960s and early 1970s the 'Fordist' consensus (after the car manufacturer, Henry Ford) – based on economic planning, mass production, structured international financial systems, and full employment – began to disintegrate. An international division of labour was emerging based on newly industrialising countries, at the same time as widespread industrial unrest and declining productivity in developed

BOX 4.1 MODELS IN POLITICAL ECONOMY

What is capitalism?

Societies have moved through four different organising systems: primitive accumulation (bartering economies); antiquity (based on slavery); feudalism (supported by serfdom) and capitalism. Capitalism refers to a social and economic system that is divided into two classes: those owning the means of production (land, machinery and factories, etc.) and those selling labour power. Under the capitalist mode of production, labour power is exploited to provide surplus value (or profit) and capitalists compete for this profit through a system that necessitates the 'accumulation of capital'.

Marxism: a critique of political economy

Karl Marx advocated an approach to political economy that he called 'historical materialism'. This materialist concept of history captured the shifting relations between the state, the economy and society through struggles between opposites. This position was initially a critique of the model of political economy used by classical economists, such as Adam Smith and David Ricardo, whose work focused on production and exchange as somewhat isolated relationships. For Marx, a critique of political economy starts with property relations within different modes of production and then explores the relations between individuals in this context. Volume I of *Capital* took these concerns forward through an analysis of the commodity form, the nature of labour power as a commodity, the labour process, the working day and alienation under capitalism. Volume II of *Capital* discusses the role of finance and money under capitalism. Volume III of *Capital*, which was completed after Marx's death by his colleague Frederick Engels, focuses more on economic reproduction. Further volumes were planned to examine the state and other aspects of the capitalist mode of production.

Structural Marxism: a critique of classical Marxism

This was dominant in the 1960s and 1970s and had close relations with political practice, especially in France. The leading thinker in structural Marxism, Louis Althusser, challenged what he saw as the technical and economic determinism within Karl Marx's thinking. Althusser introduced non-economic levels of analysis – such as consciousness and politics – into a Marxist framework and these were critically seen as 'relatively autonomous' because they formed an 'over-determined social structure'. This approach has also been termed 'base–superstructure' analysis, where historical materialism becomes a way of tracing the connections between the main social elements. More recently social scientists, drawing on the work of US economists Stephen Resnick and David Wolff, have revisited some of this thinking. Geographers such as J.K. Gibson-Graham use the term 'anti-essentialism' to reject economic determinism of all kinds: they escape capitalism through developing anti-capitalist spatial analysis and anti-capitalist political strategies.

Neo-Marxism: rebel sons of Althusser

This mode of thinking originated, first, as a return to some of the principles of Marx's *Capital*. Authors such as Ernest Mandel and Paul Baran insisted on the necessity of rates of profit and labour theories of value as keys to studying the depressions of the 1970s. Another brand of neo-Marxism has been associated with development theory, and found in the work of André Gunder Frank. A third neo-Marxist strand can be located in critical theory associated with the Frankfurt school and more closely associated with the systems theory analyses of Jürgen Habermas and Claus Offe. A fourth strand challenges base–superstructure analysis and what is seen as the automatic reproduction of capitalism. Two distinct neo-Marxist groups argue that social action is situated within, but not reduced to, structural contexts. The social structures of accumulation (SSA) school seeks to explain the role of political and economic institutions in the making of capitalism. This is a North American approach found in the work of David Kotz, Michael Reich and colleagues. Another group of scholars answers this question by developing a *régulation* approach to growth and crisis, which uncovers 'mediating mechanisms' that help to bring about conflict resolutions under capitalism. Key regulationist authors are Michel Aglietta, Robert Boyer, Bob Jessop and Alain Lipietz.

Key readings: Jessop (1997b), Lipietz (1988) and Peet and Thrift (1989).

capitalist economies, and France experienced stagflation (the coexistence of unemployment and inflation). Despite attempts to resolve this, state intervention exacerbated national economic instability.

Set within this context and also reacting against the structural Marxism of the 1970s (see Box 4.1), regulationists offered an analysis of socio-economic change that tried to understand the importance of rules, norms and conventions at a number of spatial scales (local, regional, national and supranational) in the mediation of capitalism. Regulationists explore the regulation of economic life in its broadest sense, acknowledging that capitalist development does not possess its own 'self-limiting mechanisms' or follow an 'exclusive economic logic' (Aglietta 2000). Regulationists argue that socially embedded institutions and their networks, expressed as a series of 'structural forms', are crucial to the continued existence of capitalism, despite contradictions and crisis tendencies. The initial work of Michel Aglietta captured concern with the roles played by trade unions, financial institutions and, perhaps most important, the state and its changing institutional forms and functions under capitalism (Aglietta 1978).

This thinking has been extended by others through research on 'modes of regulation' (Lipietz 1988), 'modes of social regulation' (Peck and Tickell 1992) and 'social modes of economic regulation' (Jessop 1994) – terms that capture, among other things, the different institutional forms and functions of the state. For Robert Boyer (1990), the mode of regulation denotes five levels of analysis under capitalism: the wage relation, or wage–labour nexus; forms of competition and the enterprise form; the nature of money and its regulation; the state and its forms and functions; and the international regime. When these act in concert, a period of stable growth known as a 'regime of accumulation' is said to exist. Figure 4.1 depicts this relationship and Box 4.2 summarises some key terms in *régulation* theory.

Aglietta's research on the US economy between 1840 and 1970 identifies five regimes of accumulation, each associated with a particular mode of state intervention, complementary economic system and form of state territoriality. Discussing the United States after the Civil War, for instance, Aglietta talks about the importance of a territorial ideology called the 'frontier principle', which secured economic growth

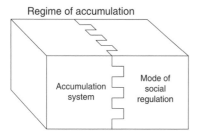

Regime of accumulation

Figure 4.1 Regime of accumulation

Source: redrawn from Peck and Tickell (1992: fig. 1), copyright 1992, with the permission of Elsevier

based primarily on agricultural production and the creation of urban commercial centres (see Chapter 2). This involved the 'domestication of geographical space' by those charged with conquering territory and building

railroads in line with a model of capitalism fostered on mobility and mutual competition (Aglietta 1978).

The *régulation* approach, then, is not restricted to Fordist analysis – a common mistake made by critics (see Brenner and Glick 1991). That said, it is more common for authors to use the example of Fordism – whose regime of accumulation can be analysed as a system supporting a virtuous model of production and consumption, and a mode of regulation consisting of: labour relations fostered on collective bargaining; the nationalisation of monopolistic enterprises; the creation and maintenance of national money; and the dominance of the Keynesian welfare state (Jessop 1992). Fordism also had a particular spatial pattern, or 'mode of societalisation', and links are frequently made between mass production and large-scale urbanisation in North America and Western Europe (Esser

BOX 4.2 THE REGULATIONIST VOCABULARY

Regime of Accumulation (RoA)

This is used to denote a coherent phase of capitalist development. There are connections here with 'long waves' of economic growth, which emphasise technological phases of development (Marshall 1987). The regime of accumulation, however, is not reduced to purely techno-economic concerns: the RoA is forged through the 'structural coupling' of accumulation and regulation, and this develops through 'chance discovery', involving trial-and-error experimentation.

Accumulation system

This explores the production–consumption relationship, whereby the individual decisions of capitalists to invest are met by demand for their 'products' through the market place. Convergence between production and its ongoing transformations and the conditions of final consumption can provide the basis for a RoA.

Mode of regulation

At one level this captures the integration of political and social relations, such as state action and the legislature, social institutions, behavioural norms and habits, and political practices. For the purposes of undertaking research, modes of regulation can be unpacked as: the wage relation; forms of competition and the enterprise system; money and its regulation; the state and its forms and functions; the international regime. The effectiveness of these institutional forms and their interrelations varies over time and across space.

Collectively these three terms allow regulationists to analyse the economy in its 'integral' sense, i.e. they are concerned with the social, cultural, and political context in which economic reproduction occurs. The spatial aspects of this are often expressed through:

Mode of societalisation

A term used to discuss the pattern of institutions and social cohesion, or the spatial patterning of regimes of accumulation.

Key readings: Jessop (1992) and Tickell and Peck (1992).

and Hirsch 1989). We discuss this further in Box 4.3. Last, given that the *régulation* approach is concerned with analysing the 'institutional infrastructure around and through which capitalism proceeds' (Tickell and Peck 1995: 363) this infrastructure varies within and between nation states, according to the different sets of economic, social and political circumstances. Geography matters and Table 4.1 highlights the key national variants of Fordism. In each case, the state has a different institutional form and performs different functions to underpin models of economic development and also instigate social and cultural change.

What comes after Fordism?

Debates within the *régulation* approach have been preoccupied with what comes after the Fordist regime of accumulation. At the other end of the spectrum, a post-Fordist camp claims that flexible specialisation or flexible accumulation is emerging. This draws on developments taking place in industrial districts across North America and Western Europe, which have a model of economic growth built on flexible small firms and specialised high-technology production (Scott 1988a, b). This is often supported by observations on localised modes of regulation. Sebastiano Brusco and Enzio Righi, for instance, draw attention to locally based institutions in Modena (north Italy), which have

helped to forge a consensus around flexible economic growth (Brusco and Righi 1989). In extreme instances, authors such as Michael Piore and Charles Sabel selectively deploy regulationist language to push these localised observations further as one-region-tells-all scenarios. Flexible specialisation is presented as an economically and socially sustainable new regime of accumulation (see Piore and Sabel 1984). Box 4.3 summarises some of the key characteristics of post-Fordism and compares these with Fordism.

At the other end of the spectrum, there are those sitting in the *after*-Fordist camp, which sees the contemporary stage of capitalism as still-in-crisis and not representing a new regime of accumulation. No prediction is made concerning the successor model to Fordism because *régulation* theory does not make any claims about the future (Peck and Tickell 1995). Their research focuses, among other things, on how the state's forms and functions have been changing in the *after*-Fordist era (see Jessop 1994; Jones 1999; Moulaert 1996). These authors criticise post-Fordist accounts for generalising from a limited number of local case studies and overemphasising the successes of this model to create sustainable and equitable growth (Amin and Robins 1990; Lovering 1990). By using case studies, the remainder of the chapter discusses the changing institutional forms and functions of the capitalist state to get behind some of these important debates.

Table 4.1 Variants of Fordism

Type of Fordist regime	Characteristics of coupling	Examples
Classic Fordism	Mass production and consumption underwritten by social democratic welfare state	United States
Flex-Fordism	Decentralised, federalised state. Close co-operation between financial and industrial capital, including facilitation of interfirm co-operation	West Germany
Blocked Fordism	Inadequate integration of financial and productive capital at the level of the nation-state. Archaic and obstructive character of working-class politics	United Kingdom
State Fordism	State plays a leading role in creation of conditions of mass production, including state control of industry. *L'état entrepreneur*	France
Permeable Fordism	Relatively unprocessed raw materials as real leaders of economy. Private collective bargaining but similar macro-economic policy and labour management relations to classic Fordism. 'Bastard Keynesianism'	Canada, Australia
Delayed Fordism	Cheap labour immediately adjacent to Fordist core. State intervention played key role in rapid industrialisation in the 1960s	Spain, Italy
Peripheral Fordism	Local assembly followed by export of Fordist goods. Heavy indebtedness. Authoritarian structures coupled with movement for democracy, attempts to emulate Fordist accumulation system in absence of corresponding MSR	Mexico, South Korea, Brazil
Primitive Taylorism	Taylorist labour processes with almost endless supply of labour. Bloody exploitation, huge extraction of surplus value. Dictatorial states and high social tension	Malaysia, Bangladesh, Philippines

Source: adapted from Tickell and Peck (1995: table 1)

BOX 4.3 FORDISM AND POST-FORDISM

Fordism and post-Fordism can be analysed under three main headings: the relations of production, the socio-institutional structure and geographical form.

Fordism

- *Relations of production.* Mass production, economies of scale, large firms and monopolistic competition, product and job standardisation.
- *The socio-institutional structure.* Collective bargaining through trade unions, demand management by the state, and mass consumption through the welfare state.
- *Geographical form.* The manufacturing belt of the United States and the zone of industrial development in Europe stretching from the Midlands of England through North West France, Belgium and Holland, to the Ruhr of Germany, with many outlying districts.

Post-Fordism

- *Relations of production.* Niche small batch production, economies of scope, small and high-tech firms, specialised products and jobs ('knowledge-based' workers).
- *The socio-institutional structure.* Individualised bargaining and decline of trade union activity, supply-side state intervention and selective consumption through welfare privatisation.
- *Geographical form.* 'New industrial spaces', such as Route 128, Silicon Valley and Orange County (North America), Baden-Württemberg (Germany), Emilia-Romagna (Italy) and Cambridge (Britain).

Key readings: Scott (1988a, 1988b) and Jessop (1992).

Régulation approaches to the state: five examples

Economic and industrial development

The relationship between the state's institutional forms and functions and the economy is discussed in research by Sean DiGovanna (1996). He analyses the roles played by 'institutionalised compromises' in the development of regions, using the *régulation* approach to compare the institutional basis and development of three economies (Emilia-Romagna in Italy, Baden-Württemberg in Germany and Silicon Valley in the United States). These regions are selected because they broadly correspond to the model of industrial districts suggested by 'flexible specialisation theory' – where industrial clustering occurs within sectors associated with electronics, aerospace and high technology in general (see Krätke 1999). DiGiovanna's research reveals differences in both the institutional foundation and the economic trajectory of each region, which result from characteristics within the mode of social regulation.

Box 4.4 summarises DiGiovanna's (1996) argument by analysing the three regions as different 'systems of regulation'. The first institutional form covers regional industrial relations and the structure of the labour market. DiGiovanna details how employers and employees relate to each other within the three regions, especially through skills development and training policies. The second institutional form cap-

tures market-based relations and forms of competition. Attention is drawn to the size of firms and sub-contracting networks. Again, relationships are different in the three regions. Baden-Württemberg is characterised by large firms, whereas Emilia-Romagna and Silicon Valley rely more heavily on small-firm alliances. The third institutional form deals with consumption regimes – such as market relations, inter-firm transactions, the different spatial structures of the firm, and product flows within the regional economy. Striking differences exists between the three regions.

Last, and perhaps most important, DiGiovanna (1996) discusses the state as an institutional form capable of intervening in the economy to provide the necessary atmosphere for economic development. The state's forms and functions are very different across the three regions. In Baden-Württemberg the state plays an important role in managing the production system and its industrial relations. In Emilia-Romagna the state's role is focused more on social reproduction and is more 'paternalistic'. Silicon Valley is an interesting example of what is becoming known as the 'knowledge-based economy', where highly skilled workers are the source of innovation and economic success. National government expenditure on research and development and on defence supports many of the organisational structures within this 'modern quicksilver economy' (Leadbeater 2000), whereas local-level government deals more with housing and environment concerns. Because California is a leading contributor to global warming, through high levels of car ownership

BOX 4.4 INSTITUTIONAL FORMS AND THE THREE INDUSTRIAL DISTRICTS

Emilia-Romagna

- *Wage–labour nexus*. Labour supply is segmented by employer and there is a clear divide between large firms (the primary sector) and small firms (the secondary sector). The primary sector is unionised and offers security, whereas the secondary sector is more precarious.
- *Forms of competition*. Competitive advantage is secured from large numbers of specialist small firms, often working together on joint marketing and technology acquisition.
- *Consumption regime*. Many firms are dependent on decisions made outside the region and few products are consumed in the region.
- *Role of the state*. Local government is socialist or communist and heavily involved in social reproduction. There is reluctance to be involved in industrial relations, and the exploitation of labour often goes unchecked.

Baden-Württemberg

- *Wage–labour nexus*. A corporatist model with high rates of unionisation in large and small firms and the determination of wages by federal and regional patterns of negotiation. Security occurs by a commitment to training and skills development.
- *Forms of competition*. Large firms dominate subcontracting relationships and lead developments in training and technology acquisition. Small firms are highly competitive and are hesitant to collaborate.
- *Consumption regime*. Components are produced by smaller firms and consumed by larger firms within the region. Suppliers often follow firms in Baden-Württemberg to foreign locations.
- *Role of the state*. The *Land* regional government is proactive in education, training and networking. The federal government is also supportive of research and development and technology diffusion. Struggles often occur between these two scales of regulation.

Silicon Valley

- *Wage–labour nexus*. Based around a bifurcated labour market model, where highly educated scientific workers operate alongside low-skilled production workers (often women and immigrants). Unionisation rates are low and violations of health and safety standards are not uncommon.
- *Forms of competition*. Dynamic strategic alliances exist between relatively small designers and customised equipment producers. This is augmented by extensive subcontracting between firms and collaboration on research, development and technical innovations.
- *Consumption regime*. Products are created largely for large and small manufacturers in worldwide markets. Many products are aimed for niche markets which are not accessed by large manufacturers.
- *Role of the state*. Local governments are not key players in Silicon Valley outside dealing with housing and environment concerns. Instead, economic development is fostered by venture capital institutions, founding firms, universities and the more informal networks of social capital.

Key reading: DiGiovanna (1996).

and traffic congestion, environmental regulation is becoming increasingly important in this region.

Geographies of scale: local modes of social regulation

An interesting application of *régulation* theory can be found in the work of Jamie Peck and Adam Tickell. This examines the state's changing forms and functions in relation to economic, political and social processes, by incorporating the geography of scale into *régulation* theory (Peck and Tickell 1992, 1995). For these geographers, spatial scales – such as regions and localities – are fluid and actively produced, as opposed to being fixed and static. Moreover, in the event of being produced, spatial scales can constrain some forms of activity and enable others to exist (see Smith 2003). We visit these themes in Chapter 6. By using the example of England's south-east region, a key space within Britain's response to globalisation, Peck and Tickell argue that political projects (in this case Margaret Thatcher's Conservative Party brand of neoliberalism) can mobilise geographical difference. They offer a regulationist reading of scale and uneven development by suggesting that modes of social regulation are mixtures of different 'regulatory systems', 'regulatory forms' and 'regulatory mechanisms' and these all operate at different spatial scales. Table 4.2 details these concerns, some of which point to the differently scaled forms and functions of the state.

Peck and Tickell suggest that 'regional couplings' occur between accumulation and regulation and this gives rise to regional or 'local modes of social regulation'. This allows them to undertake a political geography of the south-east during the late 1980s, with this region representing a particular social structure within accumulation (Peck and Tickell 1992). Figure 4.2 details the south-east standard region, which covers the Home Counties and London. Peck and Tickell argue that in order to sustain a regime of accumulation uneven development needs to be contained and their research highlights the inability of Thatcherism to control growth, such that the south-east 'bubble' burst during the early 1990s. This region's model of economic growth was fuelled by a neoliberal ideology of

'individualism' and 'ownership' (see Box 4.5), which represented a challenge to the Fordist consensus of mass production, mass consumption and one-nation social democracy. This manifested itself as the consumer credit and mortgage boom of the late 1980s that, when combined with wage inflation resulting from skill shortages and recruitment difficulties, produced an overheated economy, and rapid and uncontrollable increases in house prices. Peck and Tickell point out that these problems occurred partly because 'appropriate mechanisms for the regulation and reproduction of the economy had not been set in place' (1995: 35). The region suffered a 'regulatory deficit' and this raises questions about the sustainability of post-Fordism.

Political geographies of the local state

The work of Mark Goodwin and Joe Painter has been important for developing links between the *régulation* approach, the local state and local politics. Goodwin's research has focused on the changing institutional forms and functions of the local state and how these can act as both 'agent and obstacle' to regulation (Goodwin *et al*. 1993). Goodwin has argued that local states are products of uneven development: they have historically attempted to ameliorate the worst effects of socio-spatial polarisation by providing – through housing, education and transport, etc. – the local means for securing collective consumption (Duncan and Goodwin 1988). Building on this, Goodwin discusses the ways in which regulation, its codes and its decision-making procedures, occur not in a national territorial vacuum, but through sub-national state agencies, which 'are often the very medium through which regulatory practices are interpreted and delivered' (Goodwin *et al*. 1995: 250). Painter and Goodwin explore the notion of 'regulation as process' – where the institutional forms and functions of the state are not only seen as being associated with trying to secure stability; they are also concerned with managing fluidity, flux and change, which are 'constituted geographically' (Painter and Goodwin 1995). This emphasises the 'ebb and flow' of regulatory processes across time and space by using a 'modified version' of

Table 4.2 Regulatory forms and mechanisms at different spatial scales

Regulatory form/ mechanism	Spatial scale		
	Regional/local	*Nation-state*	*Supranational*
Business relations (including forms of competition)	Local growth coalitions	State policies on competition and monopoly	Trade frameworks
	Localised inter-firm networks	Business representative bodies and lobbying groups	Transnational joint venturing and strategic alliances
Labour relations (including wage forms)	Local labour market structures and institutions	Collective bargaining institutions	International labour and social conventions
	Institutionalisation of labour process	State labour market and training policy	Regulation of migrant labour flows
Money and finance	Regional housing markets	Fiscal structure	Supranational financial systems
	Venture capital and credit institutions	Management of money supply	Structure of global money markets
State forms	Form and structure of local state	Macro-economic policy orientation	Supranational state institutions
	Local economic policies	Degree of centralisation/ decentralisation in state structures	International trading blocs
Civil society (including politics and culture)	Local trade union/ production politics	Consumption norms	Globalisation of cultural forms
	Gendered household structures	Party politics	Global political forms

Form of sub-national uneven development

Form of international uneven development

Source: reprinted from Peck and Tickell (1992), copyright 1992, with the permission of Elsevier

régulation theory that can explore the plethora of new institutions emerging in the local state. New institutions often incorporate business sector elites to undertake the delivery of economic development and have a specific rather than a multi-functional policy remit, operating through territories smaller than those of local government (Peck 1995). Set within the context of a neoliberalist shift from 'managerialism' to 'entrepreneurialism' (Harvey 1989b), local authorities – key players under the Fordist mode of regulation and underwriters of many of its consumption norms – now operate within a system called 'local governance'. As

Figure 4.2 The south-east standard region

Source: adapted from Allen *et al.* (1998: map 2.1)

BOX 4.5 CHARACTERISTICS OF NEOLIBERALISM

Neoliberalism is a political philosophy stressing six central concerns:

1 *Liberalisation.* Promoting the free market.
2 *Deregulation.* Reducing central state intervention and direct control.
3 *Privatisation.* Selling off nationalised and state-controlled parts of the public sector.
4 *Re-commodification.* Packaging remaining parts of the public sector to behave on a commercial basis.

continued

5 *Internationalisation*. Stimulating globalising market forces.
6 *Individualisation*. Creating the opportunities for entrepreneurial activity within high-income earners.

Elements of this were realised in the 'new right' political strategy of *Thatcherism*, taken as the period of British political economy from 1979 to 1997 and covering the Conservative Party under the leadership of both Margaret Thatcher and John Major. According to Ray Hudson and Allan Williams, 'The Thatcherite project was above all else, an attempt radically and irrecoverable to redefine the relationships between the state, economy and society, and to break out of the old social democratic consensus of One Nation politics' (Hudson and Williams 1995: 39). These characteristics can also be applied closely to the United States, under especially the Reagan administration, and public sector restructuring in New Zealand throughout the 1990s.

Key readings: Allen *et al.* (1998), Hudson and Williams (1995), Jessop *et al.* (1988), Larner (2000), Brenner and Theodore (2002) and Brodie (1997).

we have detailed in Figure 4.3, governance captures the broader concern with how the local state is managed not only through elected local government but also through:

> central government, a range of non-elected organizations of the state (at both central and local levels) as well as institutional and individual actors from outside the formal political arena, such as voluntary organizations, private businesses and corporations, the mass media and, increasingly, supranational institutions, such as the European Union (EU). *The concept of governance focuses attention on the relations between these various actors.*
>
> (Goodwin and Painter 1996: 636, our emphasis)

Table 4.3 summarises this work on new developments taking place in local governance, which should *not* be read as applying only to Britain. New local 'sites of regulation' are common across many developed capitalist economies (see Brenner and Theodore 2002). Local governance, then, allows political geographers to present the local state as a system of regulation that involves different actors and regulatory practices: sometimes this is based on government and at other times governance is the norm. It is, therefore, not accurate to talk about a binary shift from local government to local governance. Many have made this mistake (see the debates in Valler *et al.* 2000) and have fallen into the same trap as those offering post-Fordist forms of analysis (see p. 61 above). Governance present does not presuppose government past. Furthermore, developments need to be related to processes occurring *outside* the local state to assess the effectiveness of the new institutions. Painter and Goodwin use the term 'local regulatory capacity' to probe on such issues, discuss the impact of these shifts within Sunderland (in the north-east of England) and claim that there is little evidence of new institutional state forms and their functions providing the necessary mechanisms for stabilising a new mode of regulation (see Box 4.6). Sunderland has a 'deficit in local regulatory capacity' and some state forms and functions are 'clearly counter-regulatory' (Painter and Goodwin 2000). This last point has been explored further by others, who suggest that the state forms and functions become modified to deal with policy problems created by previous rounds of state intervention (Jones and Ward 2002, forthcoming). The German state theorist Claus Offe called this situation the 'crisis of crisis-management' (Offe 1984) and in some circumstances this raises issues of political legitimacy that can ultimately threaten the state's operation. This is discussed further in our next case study.

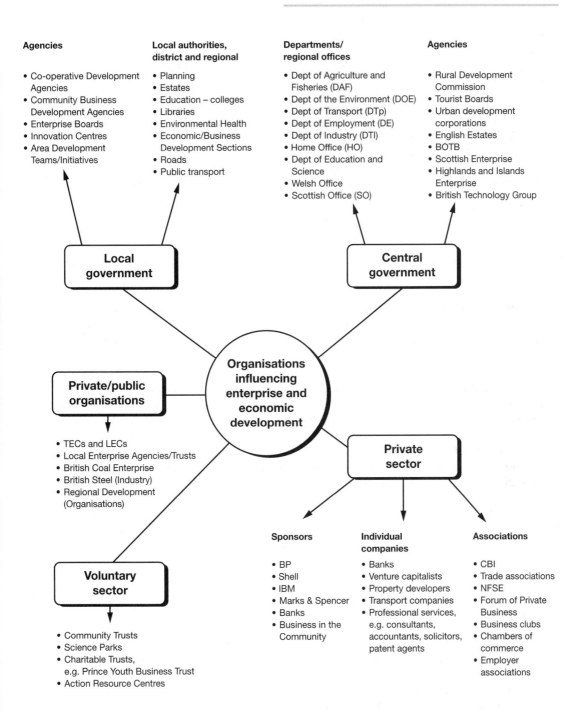

Agencies

- Co-operative Development Agencies
- Community Business Development Agencies
- Enterprise Boards
- Innovation Centres
- Area Development Teams/Initiatives

Local authorities, district and regional

- Planning
- Estates
- Education – colleges
- Libraries
- Environmental Health
- Economic/Business Development Sections
- Roads
- Public transport

Departments/ regional offices

- Dept of Agriculture and Fisheries (DAF)
- Dept of the Environment (DOE)
- Dept of Transport (DTp)
- Dept of Employment (DE)
- Dept of Industry (DTI)
- Home Office (HO)
- Dept of Education and Science
- Welsh Office
- Scottish Office (SO)

Agencies

- Rural Development Commission
- Tourist Boards
- Urban development corporations
- English Estates
- BOTB
- Scottish Enterprise
- Highlands and Islands Enterprise
- British Technology Group

Local government

Central government

Private/public organisations

Organisations influencing enterprise and economic development

- TECs and LECs
- Local Enterprise Agencies/Trusts
- British Coal Enterprise
- British Steel (Industry)
- Regional Development (Organisations)

Private sector

Voluntary sector

- Community Trusts
- Science Parks
- Charitable Trusts, e.g. Prince Youth Business Trust
- Action Resource Centres

Sponsors

- BP
- Shell
- IBM
- Marks & Spencer
- Banks
- Business in the Community

Individual companies

- Banks
- Venture capitalists
- Property developers
- Transport companies
- Professional services, e.g. consultants, accountants, solicitors, patent agents

Associations

- CBI
- Trade associations
- NFSE
- Forum of Private Business
- Business clubs
- Chambers of commerce
- Employer associations

Figure 4.3 Actors involved in promoting local economic development

Source: redrawn from Richardson and Turok (1992: fig. 2.1)

Table 4.3 New developments in local governance

Sites of regulation	Local governance in Fordism	New developments
Financial regime	Keynesian	Monetarist
Organisational structure of local governance	Centralised service delivery authorities Pre-eminence of formal, elected local government	Wide variety of service providers Multiplicity of agencies of local governance
Management	Hierarchical Centralised Bureaucratic	Devolved 'Flat' hierarchies Performance-driven
Local labour markets	Regulated Segmented by skill	Deregulated Dual labour market
Labour process	Technologically undeveloped Labour-intensive Productivity increases difficult	Technologically dynamic (information based) Capital-intensive Productivity increases possible
Labour relations	Collectivised National bargaining Regulated	Individualised Local and individual bargaining 'Flexible'
Forms of consumption	Universal Collective rights	Targeted Individualised contracts
Nature of services provided	To meet local needs Expandable	To meet statutory obligations Constrained
Ideology	Social democratic	Neoliberal
Key discourse	Technocratic/managerialist	Entrepreneurial/enabling
Political form	Corporatist	Neocorporatist (labour excluded)
Economic goals	Promotion of full employment Economic modernisation based on technical advance and public investment	Promotion of private profit Economic modernisation based on low-wage, low-skill, 'flexible' economy
Social goals	Progressive redistribution/social justice	Privatised consumption/active citizenry

Source: adapted from Goodwin and Painter (1996: table 2)

Urban politics, citizenship and legitimacy

This work on the local state's changing functions raises important issues of urban politics, especially in relation to citizenship. Ade Kearns, for instance, talks about the consequences of the shift towards multi-agency modes of delivery in the local state for 'senses of belonging to a community that lies at the heart of existential citizenship' (Kearns 1995: 169). Under Fordism, local government was important in creating 'certainty' and 'clarity': it was the main regulatory mechanism operating within the local state. With the arrival of non-elected local agencies, often drawing their personnel from outside the locality and driven by service agreements and market ethics, Kearns highlights the emergence of 'confused citizenry' and somewhat diluted senses of place (1995: 169). For Kearns, fundamental tensions exist in neoliberal local governance between service-based and citizen-oriented

BOX 4.6 'LOCAL REGULATORY CAPACITY' IN SUNDERLAND

Fordism in Sunderland

Sunderland represents the model of 'blocked Fordism' as outlined by Tickell and Peck in Table 4.1. This locality in England's north-east – one of the five districts that constituted the former metropolitan county of Tyne and Wear – was dominated by heavy industry (mainly shipbuilding and coal), had a dominance of male full-time employment, was regulated by high levels of unionised workers and was underwritten by a welfare state system that provided a social wage. Fordist local governance also provided large-scale public housing, further supporting this model of economic growth. During the era of state modernisation in the 1960s the Washington New Town introduced interventionist planning and considerable central government resources, which temporarily absorbed the crisis tendencies of Fordism.

Sunderland and the crisis of Fordism

As with most resource-based regional economies, Sunderland went through a period of intense economic restructuring during the 1970s. The heavy industry of the past gradually disappeared and was replaced by an expanding service sector. Washington New Town attracted small-scale manufacturing that offered low-skill work. By the early 1990s two-fifths of the male population of working age had no direct income from a job, whereas half of women worked in low-skill and often part-time jobs. Sunderland was very much becoming an industrial wasteland.

Is Sunderland post-Fordist?

During the 1990s Sunderland experienced an entrepreneurial city council with a shift from single-agency approaches to partnership-based organisations. This had the potential to provide the institutional basis for a sustained period of post-Fordist growth. The Tyne and Wear Development Corporation, Sunderland City Challenge, Sunderland City Training and Enterprise Council, the City of Sunderland Partnership and Sunderland Business Link were the key players in economic development and were largely involved in supply-side interventions to promote the locality to inward investors and increase the skills of the unemployed in a shrinking labour market. These institutions promoted Sunderland as 'the advanced manufacturing centre of the north', but without co-ordinated demand-side intervention policies these institutions did not possess the 'regulatory capacity' to intervene in the locality and regulate the contradictions of after-Fordism. The local state is driven by agendas and funding regimes determined outside the locality. The City of Sunderland Council and the various partnership organisations also have little impact on wage relations and the norms of collective consumption: they dealt with firefighting the consequences of after-Fordist economic decline, rather than paving the way for post-Fordist high-tech prosperity.

Key reading: Painter and Goodwin (2000).

strategies (Kearns 1992, 1995). We discuss this further in Chapter 8.

Using the example of Los Angeles, Purcell's work also considers the complex links between the changing forms and functions of the state, citizenship and political legitimacy (see Purcell 2001, 2002). Working within the *régulation* approach, Purcell presents a 'consciously political conception' of the state to draw attention to bottom-up state–citizen relations that can challenge the state's functions. Although the state has ultimate political authority within its given territory (see Chapter 2), this is derived from the collective willingness of its citizens to be ruled. The state, then, is involved in a careful balancing act, set around what Purcell calls 'mutual expectations': citizens expect the state to meet, or at least to perceive that it can meet, its obligations in return for territorial allegiance.

Purcell demonstrates this through research in Los Angeles and Box 4.7 summarises how the state changes its interventions in response to state–citizen tensions. Los Angeles (LA) contains 3.5 million people and is an incubator for 'secession movements' – groups which have turned their back on mainstream political parties and prefer to pursue more unconventional ways of making themselves heard (Purcell 2001, 2002). Purcell gives examples of such movements in the San Fernando valley, a 'microcosm of the twentieth-century suburban America' (Purcell 2001: 617), focusing on an organisation called Valley VOTE (Voters Organized Together for Empowerment). Also by drawing on the Staples Center project – a previously run-down area of LA that is now the 'entertainment centre of the world' and home to basketball, ice hockey and football teams – Purcell explores the tensions between the

---

BOX 4.7 CRISES OF LEGITIMACY IN THE LOS ANGELES LOCAL STATE

VOTE is a coalition of valley business interests and valley home owners' groups. Both are influential forces: in some cases home-owner groups have 2,000 members. Although there are conflicting agendas within the coalition – with home owners wanting controlled growth and development and the business community advocating *laissez-faire* land use policy and low taxes – they agree on key reasons for secession. Both sides feel that the City of Los Angeles (local government) is too large to be responsive to local needs and has been 'short-changing' the valley in terms of providing the necessary level of services for collective consumption. Interesting examples here are struggles over the ownership of water, with the urban population being privileged over suburban interests in the valley region. After many struggles the City of Los Angeles was forced to launch an independent charter reform commission, which recommended a rewriting of the charter for public services to defuse organisations such as VOTE.

In the example of the Staples Center Project a local commercial real estate agent, who at the same time was an adviser to the mayor, was offering tax breaks and nominal rents deals (worth $70 million and 25 per cent of the arena's estimated cost) to attract developers to this area. The rather clandestine processes at work here were uncovered by a populist councillor, Joel Wachs, who started a group Citizens Against Secret Handouts (CASH). This argued that the mayor's office should offer less costly incentives to develop the site. Because his concerns were not taken seriously within city government, Wachs balloted citizens in a move that would require a voter referendum, enforced by city law, for any new sports facility. An anti-arena movement quickly developed and it appeared that Wachs's efforts would derail the project. The developers avoided this by renegotiating a deal whereby they would absorb the costs of the project, with minimal costs being picked up by the city government. The 'city chose to assuage the discontent among its citizens rather than meet the imperatives of economic development' (Purcell 2001: 308).

Key readings: Purcell (2001, 2002) and http://www.secession.net/.

'competition state' (Cerny 1997), focused on economic competitiveness strategies such as inward investment and the maintaining of political legitimacy. The politics of economic development thus relates to defending a form of growth *and* preserving state–citizen relations (Purcell 2002).

Towards workfare states

Since the middle of the twentieth century the welfare state has dominated the political landscape of North America and Western Europe. During the 1980s and 1990s, and set firmly within the context of globalisation, advanced nations have been addressing the problems associated with economic decline and spiralling public expenditure by restructuring the institutional forms and policy functions of the welfare state. For *régulation* theorists, welfare state restructuring entails the displacement of 'passive' with 'active' forms of labour market regulation (see Jessop 2002; Peck 2001). Within active labour market regulation, the 'work ethic' is being used to reconfigure the universal rights and needs-based entitlements to welfare that characterised the state's historical commitment to full employment and social rights for all citizens. The term 'workfare' – literally meaning welfare + work – is becoming increasingly dominant in the political vocabulary as a means of securing a 'new paternalist' relationship between the state and its subjects (Mead 1997). Workfare introduces strict behavioural requirements and new social responsibilities to encourage the unemployed to become more employable and job-ready through compulsory participation in training and education programmes. The workplace is also presented as the best means of avoiding poverty through slogans such as 'I fight poverty. I work'. Workfare is frequently legitimised as a reaction to economic globalisation (see Chapter 3) through the need to secure labour market flexibility as the basis of competitiveness and based on these changes to the state's functions, Jessop suggests that we are moving from a Keynesian welfare state to a Schumpeterian workfare state. This also has implications for the state's institutional form (see Box 4.8).

Workfare began in the buoyant labour markets of North America (such as California), where since the 1960s state governments have been experimenting with mandatory work and training programmes to reduce the welfare case load. This increased throughout the 1970s and the Federal Family Support Act 1988 required state governments to provide mandatory work or training activities for welfare recipients as the condition of receiving benefits. Based on local success stories – such as Riverside, California (see Box 4.9) – workfare was claimed to be a national success and was transferred across North America through think-tanks and political advisors (Peck 2001). Workfare became intensified under the Personal Responsibility and Work Opportunity Reconciliation Act, which replaced the 60-year-old Aid to Families with Dependent Children programme (AFDC) with block-granted welfare payments to the state level and also introduced a time-limited unemployment benefit system. Signing this Act in August 1996, President Bill Clinton famously argued it was about 'ending welfare as we know it'. Critics highlight the long-term impact that the strategy will have on the plight of welfare children and the deepening of America's economic problems (Baratz and White 1996).

Welfare state restructuring in North America has not gone unchallenged at the local level and new spaces have been opened up for contesting state strategy through political activism within civil society. Organisations such as Workfairness and Community Voices Heard in New York have been lobbying since the mid-1990s for regulatory standards to minimise the exploitation of labour and to get workfare workers unionised. Box 4.10 describes these organisations, Figure 4.4 details a Workfairness anti-workfare leaflet, and Plate 4.1 is a protest against workfare in New York City during 1999.

These developments are not isolated to North America. Ivar Lødemel and Heather Trickey (2000) and the OECD (1999) highlight similar trends occurring across Western Europe, but in doing so reveal subtle differences in the changing institutional form and function of the welfare state. Authors such as Gøsta Esping-Andersen (1990) and Evelyne Huber and John Stephens (2001) attribute geographical differences to the role of different interest groups that can influence

BOX 4.8 FROM KEYNESIAN WELFARE STATES TO SCHUMPETERIAN WORKFARE STATES

Jessop highlights a new era for the institutional forms and functions of the state that is associated with the shift from Keynesian welfare to Schumpeterian workfare. In more recent work this is expressed as a movement from Keynesian welfare national states (KWNS) to Schumpeterian workfare postnational regimes (SWPR).

Keynesian welfare states

After the British political economist John Maynard Keynes:

- *Function of the state.* The KWNS supported full employment through demand management, provided public infrastructure to support mass production and consumption, and ensured mass consumption through collective bargaining and the expansion of welfare rights.
- *Form of the state.* The national scale was used for state intervention in economic and social policy making, with local as well as central modes of delivery.

Schumpeterian workfare regimes

After the Austrian political economist Joseph Schumpeter:

- *Function of the state.* The SWPR supports supply-side innovation and competitiveness through promoting open economies and subordinates social policy to the needs of competitiveness by pushing wages down and promoting low-skill employment.
- *Form of the state.* The national scale is no longer the dominant scale for state intervention, with the emergence of devolved local and regional 'partnerships' and networks.

Key readings: Jessop (1994, 2002).

BOX 4.9 BORN IN THE USA: WORKFARE IN RIVERSIDE, CALIFORNIA

The Riverside Greater Avenues for Independence (GAIN) model concentrates on moving welfare recipients into work as rapidly as possible and with minimum costs. Evaluations have shown it to be successful in driving costs down and accelerating the process of labour market re-entry, although there is no evidence that it can lift participants out of poverty. The Riverside model has excited much interest across North America and Western

Europe because of its 'pure' workfare appeals. It is a no-frills, high-volume, low-cost way of enforcing work participation and work disciplines. Peck (1998) uncovers a 'new mode of labour discipline' that seeks to conscript the poor into low-wage, or contingent, work. He discusses the various strategies used by officials in Riverside and draws attention to the consequences that these have on local labour markets.

Key reading: Peck (1998).

BOX 4.10 RESISTING WORKFARE

Workfairness

Workfairness is a New York-based organisation of workfare workers and their supporters that emerged as a response to New York's Work Experience Program (WEP). Workfairness has been campaigning for a better deal for workfare workers and has challenged the 'new paternalist' ideology of blaming the unemployed for their position in society. 'People on welfare have been stereotyped, maligned in the media, and made into scapegoats for the politicians, the rich and powerful to target. The truth is that Workfare mothers get up in the morning just like any other worker, they see their children are cared for, and they go to work. Workfare workers work very hard, and they are proud of the work they do. They don't want to be cheap replacements for their friends and neighbors fortunate enough to have union wage jobs. Workfare workers want permanent jobs at union wages. They want to join unions. They want respect, dignity and equality. These are the things that WORKFAIRNESS and others are trying to fight and win.'

Community Voices Heard

Community Voices Heard (CVH) is an organisation of low-income people, mostly women on welfare, working together to improve the lives of the poor in New York City. It is run by low-income people on welfare. CVH uses public education, public policy research, community organising, leadership development, political education and direct action issue organising to campaign around issues such as 'welfare activism'. In the late 1990s CVH lobbied New York City politicians to ensure that welfare reform moved people out of poverty by creating jobs, job training, education and child care. CVH has also been developing grass-roots leadership among women on welfare to recognise their power and potential to impact public policies that impact on their daily lives.

Key readings: Holmes and Ettinger (1997), http://www.iacenter.org/workfare.htm and http://www.cvhaction.org/.

DON'T CUT
OUR
FOOD STAMPS

RESTORE EMERGENCY ASSISTANCE TO PEOPLE ON PUBLIC ASSISTANCE
REAL JOBS NOT WORKFARE !

COME TO THE **PROTEST**

ON THE STEPS OF **CITY HALL**
FRIDAY APRIL 16th at 12:00 PM
(Take the east-side trains to the City Hall stop, and west side trains to Chambers St.)
Join our supporters, speak up for yourself, *IT'S YOUR RIGHT*

Mayor Giuliani is cutting all food stamps to people on public assistance between the ages of 18 and 50 after three months beginning April 1, 1999. This cruel act will deprive more than 25,000 people of food. To make matters worse, Giuliani is also ending emergency assistance to poor people who are evicted. City Hall refuses to support a jobs bill for WEP workers and people on public assistance.

The mayor can reverse the food stamp cut off, the elimination of emergency assistance to people in need, and sign legislation creating jobs instead of slave labor workfare. Come to the City Hall Protest to demand that Giuliani **STOP TREATING PEOPLE LIKE DIRT**!

For more information, or if you need assistance to get to City Hall on April 16, call:
WORKFAIRNESS *an organization of WEP workers, people on public assistance, and there supporters* - 39 West 14th St. #206, NY, NY 10011 **Phone (212) 633-6646** Fax (212) 633-2889

Union Labor Donated

Figure 4.4 Workfairness leaflet
Source: reprinted by courtesy of Workfairness, New York

Plate 4.1   Anti-workfare protest, New York City, April 1999

Courtesy of Martin Jones

those holding power within the institutions of the welfare state.

The arrival of the 'new paternalism' in the United Kingdom, for instance, has been more recent. Owing to trade union pressure, workfare was resisted during the 1980s and the post-war labour market settlement remained more or less intact until the 1996 Jobseekers' Allowance required a strict 'agreement' between the 'job seeker' and the state as a condition for the receipt of benefits (Jones 1996). With the arrival of the New Labour government in 1997, elements of workfare – presented as 'welfare-to-work' – have been clearly evident in the various 'New Deal' programmes. These require participation in a series of 'options' in return for welfare benefits. Priority is given to immediate placement in the labour market to embed the work ethic at the earliest opportunity (Peck 2001). The Chancellor of the Exchequer famously argued that 'Rights go hand in hand with responsibilities, and for young people offered new responsibilities . . . there will be no third option of simply staying at home on full benefit doing nothing' (Brown 1998: 1).

In contrast to the neoliberal approaches of the United States and United Kingdom, research undertaken on Denmark has suggested that a welfare-*through*-work strategy is being deployed and this retains many of the state's welfarist labour market functions (see Box 4.11).

BOX 4.11 WELFARE STATE RESTRUCTURING IN DENMARK

Denmark has adopted a 'welfare-*through*-work' model. Owing to the power of the labour movement and the pressures exerted against the state by gender movements, this retains some key social policy functions. First, social partnerships have been strengthened in respect of the delivery and implementation of welfare. Second, financial planning and decision making have been decentralised to regional institutions. Third, the unemployed have been given the right to counselling, an individual action plan and, more important, access to a comprehensive package of leave schemes, including job training, education and child care. A key aspect of this model is a work-sharing scheme called 'job rotation', whereby the unemployed are recruited and given direct job training experience in posts vacated by (predominantly) unskilled workers, who in turn are given the opportunity to update their training and education. The unemployed receive both work experience at trade union negotiated rates and additional vocational training. This benefits the firm by providing the sustainable basis for an up-skilled work force, without loss in employment. This initiative is being promoted in fourteen different countries through an EU-funded transnational programme.

Key readings: Etherington and Jones (2004) and Torfing (1999).

Summary

This chapter has provided an overview of the uses made by the *régulation* approach to capture the changing institutional forms and functions of the capitalist state. This approach to political economy is an on-going method and should not be read as fully finished or complete. As Aglietta puts it, 'We must speak of an approach rather than a theory. What has gained acceptance is not a body of fully refined concepts but a research programme' (Aglietta 2000: 388). The *régulation* approach doesn't have all the answers but it asks some interesting political geography questions. Tickell and Peck have highlighted five important 'missing links' – more work on modes of social regulation; more research on leading-edge motors of growth; consideration of how and why economies change; more attention to spatial scales of analysis; heightened consideration of consumption issues (Tickell and Peck 1992) – some of which are important concerns for political geographers. Jessop has also suggested that the *régulation* approach needs to be combined with other approaches, such as state theory and discourse analysis, to get a better handle on political economy (Jessop 1995).

Political geography students might wish to consider a recurrent criticism levelled against this approach to political economy, the regulationist defence and a possible extension of *régulation* theory. It is often stated that *régulation* theory tends to insert a divide between the economy, which is bracketed as a black box and simultaneously cast as a key protagonist, and the cultural and political realms (see Graham 1992; Lee 1995). In reply, regulationists claim that the economy is constructed, reconstructed and institutionalised *through* social, cultural and political relations (Bakshi *et al*. 1995). For uncovering further the changing institutional forms and functions of the state, it is also suggested that mileage can be gained from developing a 'regulationist state theory' which draws on Bob Jessop's work on states as mediums and outcomes of territorially distinct political strategies and policy projects (Jessop 1997c). We discuss this further in Chapter 9 on public policy and political geography.

Further reading

The subjects covered in this chapter are wide-ranging and there are several avenues of further reading. The political economy approach is discussed further in Castree, 'Envisioning capitalism: geography and the renewal of Marxian political economy', *Transactions of the Institute of British Geographers*, 24 (1999a), 137–58; Gibson-Graham, *The End of Capitalism (as we knew it)* (1996); Lipietz, 'Reflections on a tale: the Marxist foundations of the concepts of accumulation and regulation', *Studies in Political Economy*, 26 (1988), 7–36; Peet and Thrift, 'Political economy and human geography' in Peet and Thrift (eds), *New Models in Geography* (1989).

The papers by Tickell and Peck, 'Accumulation, regulation and the geographies of post-Fordism: missing links in regulationist research', *Progress in Human Geography*, 16 (1992), 190–218; MacLeod 'Globalising Parisian thought-waves: recent advances in the study of social regulation, politics, discourse and space', *Progress in Human Geography* 21 (1997), 530–53; Painter and Goodwin, 'Local governance and concrete research: investigating the uneven development of regulation', *Economy and Society* 24 (1995), 334–56; and Valler *et al.*, 'Local governance and local business interests: a critical review', *Progress in Human Geography*, 24 (2000), 409–28, all provide very good overviews of the development of *régulation* theory and its applications in human geography.

Those wanting to trace the intellectual origins of *régulation* theory should consult Boyer, *The Regulation School* (1990); Aglietta, *A Theory of Capitalist Regulation* (2000); Boyer and Saillard, Régulation *Theory* (2002), and two key papers by Jessop, 'Twenty years of the (Parisian) regulation approach: the paradox of success and failure at home and abroad', *New Political Economy*, 2 (1997a), 503–26, and his survey article 'The regulation approach', *Journal of Political Philosophy*, 3 (1997b), 287–326.

The debates on Fordism and Post-Fordism are covered in the excellent book edited by Amin, *Post-Fordism: A Reader* (1994); the collection of essays brought together in Storper and Scott, *Pathways to Industrialization and Regional Development* (1992); Harvey, *The Condition of Postmodernity* (1989a), and papers by Moulaert and Swyngedouw, 'Survey 15: a regulationist approach to the geography of flexible production systems', *Environment and Planning D: Society and Space* volume 7 (1989), 327–45; and Michael Webber, 'The contemporary transition', *Environment and Planning D: Society and Space*, 9 (1991), 165–82. Some of these debates have been revisited within the context of neoliberalism in a collection of essays edited by Brenner and Theodore, *Spaces of Neoliberalism* (2002).

The majority of our case studies cover the broad area of local and regional economic development, and those wishing to further explore the comparative nature of this work should consult a number of excellent sources: Harvey, 'From managerialism to entrepreneurialism: the transformation of urban governance in late capitalism', *Geografiska Annaler*, 71B (1989b), 3–17; Clarke and Gaile, *The Work of Cities* (1998); Lauria (ed.), *Reconstructing Urban Regime Theory* (1997); Walzer and Jacobs (eds), *Public–Private Partnerships for Local Economic Development* (1998); Leitner, 'Cities in pursuit of economic growth', *Political Geography Quarterly*, 9 (1990), 146–70; Wood, 'Making sense of entrepreneurialism', *Scottish Geographical Magazine*, 114 (1998), 120–3; and John, *Local Governance in Western Europe* (2001).

POLITICS, POWER AND PLACE

The political geographies
of the nation

Introduction

I was walking down this street in Mold, north Wales [in the UK] – years ago now – with some friends from my home town in Llanelli in south Wales. We were laughing, playing about and talking to each other in Welsh. A group of locals came down the street, and after hearing our accents, came up to us quite aggressively . . . obviously looking for a fight. They tried to taunt us by calling us 'Cymry plastic' or 'plastic Welsh'. Obviously, we didn't fit in with their ideal type of Welsh person. We weren't local and we spoke south-Walian Welsh. That was enough for them. Anyway, we ran into the nearest pub and managed to get out of any trouble.

This tale, narrated by Rhys Jones, one of the authors of this book, helps to illustrate a number of crucial themes relating to the ideas of place and nation. First, it demonstrates the close relationship that exists between place and nation. In this story, certain places can be seen to represent the Welsh nation more effectively than others: north rather than south Wales, Mold more so than Llanelli. It shows the way in which places help symbolise and anchor national identity. Second, the story shows us the subdivisions that exist within the allegedly homogeneous entities of nations. We usually think of nations as coherent 'imagined communities' (Anderson 1983) of people who follow the same customs and speak the same language. Indeed, this is one of the main ideological foundations of nationalism: to encourage us to believe

that it is possible to draw boundaries around homogeneous groupings of people. Rhys Jones's experience in north Wales would seem to undermine this notion. Even though he and his friends were born in the same country, and spoke the same language, as the other group of men, they were still thought of as being somehow different in nature. In this example, these imagined boundaries between the two groups were constructed along place-based and linguistic lines. To be born in a different region or a different town, or to speak a different dialect, led to the construction of imaginary boundaries between the two groups of people. Finally, the tale illustrates the close link that often exists between language and national identity. The ability or inability to speak a language can often be used as a means of defining who acually belongs to, and who is excluded from, a nation. In this case, the ability to speak a language was not deemed to be enough of a badge of national identity, since the Welsh language had to be spoken in a particular way in order to gain membership of the Welsh nation.

These are all key themes that need to be explored when discussing the link between place and nation, and we shall focus on each in turn during the course of this chapter. The first section will introduce the different kinds of theories that have tried to explain the formation of nations. Following on from this, we then proceed to explore the significance of the key geographical concepts of place, landscape and territory for the nation. Finally we focus on the contestation of nations – focusing on ideas of gender, region and localities – and thereby seek to illustrate the problematic nature of nationalism as ideology.

Reproducing nations

So what do we mean by a nation? Similar to comments made regarding the state in Chapter 2, it is often the case that our membership of a nation – as individuals and as communities of people – makes it difficult to think about it in an objective way. This is even more of an issue since it is nigh-on impossible for any individual to escape from the ideological grasp of a given nation. As Gellner (1983: 6) has noted, it is as if 'a man [*sic*] must have a nationality as he must have a nose and two ears'. When nations are commonly viewed as 'natural' phenomena like this, it becomes difficult to critically analyse their form and function. A further problem arises when attempting to distinguish between states and nations. In the first part of this section, our aim is to define what we mean by nations and emphasise the difference between states and nations.

According to Anthony Smith (1991: 14), a prominent writer on themes of nationalism, a *nation* should be viewed as a 'named human population sharing an historic territory, common myths and historical memories, a mass, public culture, a common economy and common legal rights and duties for all members'. Although definitions of nations display slight variations in the themes that they emphasise (see below), on the whole they follow the general principles laid down by Smith. Smith's definition draws our attention to a number of important themes. First, and crucially, all nations possess a geographical referent in their claims to a particular territory. It is difficult to imagine a nation that does not claim access or control over a certain territory, and it is this feature of nations that acts as the main justification for studying nations from a geographical perspective. We will argue in subsequent sections that territory is only one geographical theme that is of importance to nations. Others, such as place and landscape, are also highly significant. Second, Smith's definition stresses some of the key cultural aspects of nations. Nations possess a common 'public culture'. They also emphasise common historical memories that help to engender a sense of loyalty towards the nation. It is these common cultural characteristics that enable the members of a

nation to imagine the existence of this large-scale, yet close-knit, community of people, even though they will never meet all other members of the nation (Anderson 1983). It is these cultural themes that also help us to distinguish between states and nations. Whereas states are organisations that seek to control a particular territory – similar to nations – they do so in a relatively impersonal manner. States, *per se*, do not attempt to stress the cultural commonalities that exist within and between the state's citizens. Nations, on the other hand, are communities that emphasise common ties of custom, history and culture as a means of gaining control over a certain territory. It is when both these institutions combine, of course, that the powerful organisation of the nation-state is formed. Third, Smith emphasises the role that certain legal and economic processes play in forging a nation. Once again, this draws us near to definitions of the state (see Chapter 2) and helps to illustrate once again some of the common elements that characterise states and nations.

If nations are common communities of people that share certain cultural attributes and a particular territory, then we need to think of *nationalism* as an ideology that seeks to promote the existence of nations within the world. Furthermore, an important element within nationalism is the belief that every nation should possess its own sovereign territory or state. The ideology and political practice of nationalism therefore seek the ideal political and territorial scenario of the *nation-state*, in which every citizen of the state is a member of the same nation. Thinking of this in geographical terms, nation-states represent political geographies in which the boundary of the nation coincides with the boundary of the state (Gellner 1983: 1). Obviously, in a world characterised by continuous flows of people, the ideal of the nation-state is precisely that – an ideal that can never be achieved. Indeed, it has been famously argued by Mikesell (1983) that the only example of a nation-state in the contemporary world is Iceland. Unfortunately, the difficulties in achieving the goal of the nation-state do not stop states, nations and minority groups from trying – sometimes peaceably and sometimes violently – to reach the ideal. Examples such as the attempts to create an independent Quebec through referendums and the terrorism of

ETA, the Basque separatist movement in Spain, demonstrate the salience of such processes within contemporary political geography (see Anderson 1995).

We now possess an understanding of what nations are, along with some of the ideologies that are linked with them. The key question that has exercised the minds of social scientists and historians in the field of nationalism is 'How are nations formed and continue to exist?' To put in another way, 'How are nations reproduced?' Theories that have sought to explain the formation of nations are numerous, and draw on a broad range of political, economic and cultural viewpoints. At a general level we can conceive of them in two broad categories (see Table 5.1).

In the first category are those theories that see nations as communities of people who have some discernible roots in the pre-modern period (in other words, before approximately 1500). These theories often posit a link between nations and ethnic communities of people. Most problematic in this category are those so-called primordialist theories of the nation that argue that nations have always existed, from time immemorial, as it were. Nations are therefore not produced, as such, since they are seen to represent essential qualities – linked with race, blood, language, religion or ethnicity – that are inherent in communities of people. These can appear as quite naive interpretations of nations, since they posit that certain definable and homogeneous communities of people have always existed in the world. Halvdan Koht (1947:

278), for instance, perceived a late twelfth-century uprising in Italy as an early display of Italian nationalism against the Germanic people who were occupying their country. Koht argued that the Italians succeeded in rallying their people by proclaiming that the German language was similar to the 'barking of dogs and the croaking of frogs'. Clearly, here we see an attempt to view a medieval conflict as an early instance of nationalism. It is doubtful whether such is the case, since it is unlikely that the communities of people who lived in Italy at the time displayed the characteristics common in modern nations (though see Reynolds 1984). More seriously, these primordialist theories are also potentially dangerous ideas, since they can be used to legitimise reactionary and racist attitudes towards 'outside' people and influences that allegedly undermine the 'purity' and 'integrity' of the nation. Many instances of historical and contemporary nationalism – especially during times of conflict – have witnessed exclusionary and essentialist attitudes towards people who do not 'fit in' with the norm of the nation. In recent times, we can think of ethnic cleansing in the former Yugoslavia as a particularly repugnant example of attempts to rid the nation of 'impure' or 'undesirable' elements (e.g. Mirkovic 1996).

Less problematic are the so-called perennialist theories promoted by academics such as Anthony Smith (see especially 1986). In his attempt to chart the ethnic origins of nations, Smith has argued that

Table 5.1 Theorising the reproduction of the nation

Theorising the roots of nations in pre-modernity
- *Primordialist theories*: the nation is an essential element of the human make-up and has always existed
- *Perennialist theories*: the nation is a product of modernity but has its roots in earlier pre-modern ethnic communities

Theorising the nation as a product of modernity
- *Nationalism as 'high culture'*: industrial development requires a literate work force and state education systems 'educate' citizens regarding nationalist ideals
- *The socio-economic development of nationalism*: 'peripheral' countries mobilise their populations, using nationalism as a means of competing against the 'core'
- *Nationalism and the state*: nations are formed as a result of the creation of state bureaucracies
- *Nationalism as ideology*: nationalism is created as a means of escaping the isolation and social disruption caused by processes within modernity

Source: after Smith (1998)

'nothing comes from the nothing' (1996: 386): in other words, it is extremely difficult to create a nation out of a group of people who do not feel any sort of communal feeling towards each other. Rather than seeing nations as totally modern fabrications (see below), Smith argues that the more successful nations have been based on early ethnic communities of people, or *ethnies*. In doing so, Smith tries to emphasise the long history of nations. The processes and institutions of modernity – industrial development, capitalism and the state – impact on ethnic communities in various ways. Some ethnic communities are dissolved as a result of these processes as the language, culture and customs of a particular *ethnie* are adopted as the norm for a number of adjacent ethnic communities. This process possessed distinctive geographical undertones, since it often involved the imposition of the language spoken in the core of the nation on its peripheral regions. We can think, for instance, of the adoption of the language of the Île de France – the area around Paris in France – as the official language of the French nation, and the subsequent attempts to dilute and extinguish the other so-called vernacular languages – Breton, German and Basque – spoken in other parts of the country. Smith's ideas certainly make us think of the long-term development of nations out of earlier ethnic communities and they can act as a powerful antidote to the more explicitly modernist theories of the formation of nations.

Modernist theories view nations as communities of people that have come into being as a result of various processes that happened in the modern period. To a greater or lesser extent, they view the nation as a new creation, actively produced by means of the processes and institutions of modernity. These theories tend to focus on the civic quality of nations. Anthony Smith (1998) has classified them into four main categories of theories. We briefly discuss each category in turn. First, some theories emphasise the role played by industrial development and state education systems in the formation of national 'high cultures'. Gellner (1983) argued in his later writings concerning the important role of state educational systems in the creation of a 'high' culture necessary for successful industrial development. In this period there was a need

for individuals who understood the same language. It was this need to create an impersonal society in which all individuals spoke the same language, were educated by the same state education system, and were to a large extent, interchangeable one with another, that explained the formation of nations. State education systems replaced 'low' cultures – meaning localised and traditional customs, languages and cultural norms – with the 'high' culture, meaning a 'standardised, education-based, literate culture' (Smith 1998: 32). These shared, literate cultures became the bases for national sentiment. We can consider empirical evidence for Gellner's theory in the context of school textbooks and the key role played by them in helping to inculcate children that they belong to a specific nation. A famous example of this process relates to the different histories portrayed in Estonian school textbooks before and after the country's independence. While Estonia was still part of the Soviet Union, the Soviet Union was portrayed favourably, since it had helped to liberate Estonia from German rule towards the end of the Second World War. After independence, however, school textbooks portrayed the Soviet Union in a negative light, seeing it as a repressive political and cultural force within Estonian history (Smith 1996). This example clearly shows the key role played by state educational systems in moulding nations. Rather than representing age-old cultures and customs, as primordialists argue, Gellner argues that nations are in fact created out of new, state-based cultures, languages and customs. As such, 'nations are functional for modern society' (Smith 1998: 35) meaning that they are indispensable to it.

Second, and to some extent linked with the first set of theories, are socio-economic models of the formation of nations. Characteristic of these types of theories is the work of Tom Nairn (1977), who views nations as political and cultural entities that are formed as a result of socio-economic processes of the world economy. In a nutshell, Nairn argues that nationalism derives from the uneven development of capitalism. By this, he means that the success of the capitalist process in certain western states after 1800 depended on their exploitation of cheap labour and resources on the periphery. Nairn views nationalism as a means by which political and cultural leaders in the periphery seek to mobilise their populations against the imperialism of the core, much in the same way as a sense of nationalism can be said to help the players in a sports team to compete against an opposing team. Efforts were made, therefore, to encourage popular forms of nationalism in the periphery. Nairn uses the spread of nationalism in Germany and Italy during the nineteenth century, as they sought to oppose the capitalist might of the United Kingdom and France, as an illustration of this process. There are weaknesses in Nairn's model, for instance his inability to explain the way in which political and cultural leaders in the periphery succeeded in promoting nationalisms within the population that they ruled. Similarly, there is firm empirical evidence that the first nationalisms appeared in the core, rather than the periphery. According to Smith (1998: 53) it appeared in England, Britain, France and America before Germany, the alleged first state to promote a sense of nationalism. We can, therefore, take only so much from Nairn's model. It offers some insight into the relationship between capitalism and nationalism but it displays too many weaknesses to be considered as a truly useful explanation of the formation of nations and nationalism.

Third, we need to consider political explanations of the formation of nations and the development of nationalist ideologies. Here, the development of the modern bureaucratic state is used as an explanation for the formation of nations. We can take the ideas of Anthony Giddens (1985) as an illustrative example of this body of work (though see also Mann 1986; Tilly 1975). For Giddens, nations can only exist with regard to states. What is key here, therefore, is the consolidation of the state as a bureaucratic organisation that extends its administrative control outwards from a core to defined boundaries (see Chapter 2). The significance of this is twofold: it brought a singular and uniform administrative rationality to all citizens of the state, and it also imposed a fixed boundary on the national community of the state. In a few rare instances, this could lead to the creation of a singular nation within the boundaries of the state, but it more often led to cultural and national reactions against bureaucratic control within particular states. Some

support for Giddens's ideas appears in the context of African states. Here, we see clearly the influence of states on national communities during the post-colonial era. Leaders of these states have tried to forge a national community to match the boundaries of their administrative control, but these boundaries have also acted as the frame of reference for secessionist and independence movements within these countries. A painful example of this process is the Democratic Republic of Congo, where a number of armed revolutionaries, especially in the eastern half of the country, are challenging the territorial integrity of the state. The main problem with Giddens's ideas is their inability to deal with more cultural forms of nationalism. Not all nationalist movements seek to create their own independent state, and may rather seek to propagate forms of nationalism that seek to improve the moral and cultural well-being of the nation. In many ways, the nationalism of the Maori people of Aotearoa/New Zealand fits into this model. Maori are more concerned with ideas of cultural well-being, socio-economic development and the protection of the natural environment than with the need to create an independent Maori state (see Pawson 1992; Bery and Kearns 1996). As such, political explanations of the nation fail to account for the more diffuse form of nationalism experienced in this context.

Fourth and finally are the ideological interpretations of Elie Kedourie (1960, 1971), who views nationalism as a system of beliefs similar to religion. Here, nationalism is viewed as a system of beliefs that is promoted by groups of intellectuals as a means of making sense of the fundamental changes that affect societies as a result of the processes of modernity. Indeed, the nation's immemorial history plays an important role in anchoring the life-worlds of individuals who are lost within the modern maelstrom. Importantly, Kedourie grounds the growth of nationalism in two specific periods and places: an early nineteenth-century Central Europe that was experiencing large-scale political and social change and Africa and Asia during the late nineteenth and twentieth centuries. The key impetus to the development of nationalism, in this regard, was the colonial process that operated in particular countries. Traditional forms of living were destroyed, and local intellectuals were marginalised from colonial bureaucracies and reacted to this process of marginalisation by engaging with ideals of nationalism that diffused from western states. For instance, Kedourie (1971: 42–3; see also Anderson 1983: 195) describes the role of the Greek intellectual, Adamantios Korais, in helping to enframe Greek history within western conceptions of nationalism. Having experienced life in revolutionary France, Korais began to comprehend Greek history as a nationalist history of the Greek people. This intellectual enabled Greeks to begin to view their place in the world as one that was structured by the new secular religion of nationalism: instead of experiencing the *anomie* and isolation of modernity, they began to view themselves as members of an enduring Greek nation. According to Kedourie, therefore, nationalism should be viewed as a political and cultural opiate (Smith 1998: 103) that enables individuals to make sense of the world in which they live.

Here, then, are the theories that have sought to explain the formation of nations as political communities and of nationalism as a political ideology. We stress that the categories Smith (1998) has imposed on this broad-ranging literature should not be viewed as definitive in nature. From the discussion above, it is clear that there are many overlaps between the various categories. None the less, adopting such a system of categorisation enables us to appreciate the variety of theoretical viewpoints concerning nations and nationalism. Perhaps the key point to note at this stage is that it is unlikely that any one theory can explain the formation of nations and nationalisms in all historical and geographical contexts. Some theories are more applicable at certain times and in certain places and, as such, it is doubtful whether we will ever find an all-encompassing theory of nationalism. In any case, much academic insight may be gained from articulating the contestations and inconsistencies that exist within this variegated body of literature.

In the following section, we turn to more geographical themes by discussing the way in which the key geographical concepts of place, landscape and territory help to inform the character of nations and nationalism.

Geographies of the nation

Since approximately the 1980s, geographers have begun to examine the significance of spatial concepts for nations and nationalism. The key contribution that acted as a clarion call for political geographers to examine nationalism from a spatial perspective was that published by Colin Williams and Anthony Smith in 1983. In their paper on the 'National construction of social space' they argued that much research needed to be carried out on the geography of nations and nationalisms. As a starting point for this project, they outlined eight different contexts in which geographical themes could inform our understanding of nationalism. These are noted in Table 5.2.

Williams and Smith's (1983) ideas are an extremely valuable way of understanding the relationship between geography and geographical concepts and nationalism. Indeed, their ideas have been taken up by a number of political geographers who have brought new insights into the study of nationalism through their focus on geographical concepts such as place, landscape and territory. It is to their work that we now turn.

Placing the nation

In this section we focus on the importance of certain places for nations. By referring to place here, we are specifically concerned with place as locality, in other words, places that occur at small scales. Even though nationalism refers to an ideology that exists at a national scale, and draws members of a nation together as one common community of people, individual nations always draw on specific places as sources of ideological nourishment.

At one level, we can think of generic places that help to sustain the political ideology of certain nations. One particularly powerful type of place is the memorials to the dead of the nation (see Plate 5.1). These are individuals who have paid the ultimate price for their loyalty to their nation. Michael Heffernan (1995), for instance, has described in detail the discussions and, indeed, conflict that revolved around the commemoration of those members of the former

Table 5.2 Space, territory and nation: eight dimensions

Dimension of national space and territory	Recent examples
Habitat The nature of the environment and the soil helps to explain the location and nature of human communities. Following on from this, the members of the nation come closer to the ideal of the nation if they live close to the soil, in other words, in rural rather than urban areas	Jewish fundamentalists try to recapture the essence of the Jewish nation by creating new rural settlements in disputed territories
Folk culture Soil and environment lead to particular customs and social norms, e.g. daily, monthly and annual rhythms of life in rural areas. These become prized as peasant virtues and folk cultures. Efforts are made to avoid the cosmopolitan culture of the cities and recapture the rural roots of the nation	Nazism and the emphasis placed within this extreme nationalist ideology on the virtue of a folk culture of the *Volk*
Scale The size or scale of nations helps to position the nation in an international league table of nations. We can link this most clearly with the attempts made by nations to expand the reach of their territorial control. By doing this, the nation increases its own prestige	This was implicit in much of geopolitics in the first half of the twentieth century, and helps to explain the 'scramble for Africa' by European nations
Location We can think here of the uneasy relationship between a nation and other neighbouring nations. This can lead to warfare, which may affect the character of the nation. Also the distance between peripheral areas of the nation and the core can lead to uneasy relationships	For instance, the uneasy relationship between India and Pakistan has led to more virulent nationalisms in the two countries. We can also think of the border community of Kashmir that lies between India and Pakistan and which has been problematic for both nations
Boundary Crucial here is the idea of finding the 'natural' frontiers of the nation and the need to get away from artificially imposed borders	This idea has been important in the development of the French nation, which has professed an explicit need to defend its natural borders
Autarchy Land is viewed as a resource deposit for the benefit of a particular nation. Any struggle for land and independence is linked with the struggle for the use of national resources, first in the context of agrarian resources but also in the context of minerals	A good example of this was the attempt by the Kikuyu to wrest land from white control in Kenya. This process is being repeated in Zimbabwe
Homeland Territory is not a neutral term for the nation. It is the national homeland, the historic root of the nation	We can think of the importance of the homeland for Jews, something that sustained them during thousands of years of exile
Nation building The process of forming and maintaining nations involves improvements carried out within the nation's territory. It is this 'infrastructure of the nation' that turns a territory into a national territory	For instance, cities, communications networks, power stations, law and educational systems are all crucial to the construction of a national territory. A fine example of this is the communication node of Grand Central Station in New York

Source: after Williams and Smith (1983)

Plate 5.1 The Arc de Triomphe and its 'eternal flame': remembering a nation's dead

Courtesy of David Henry

British Empire who died during the First World War (1914–18). After much wrangling a system was adopted whereby those who had died in the war were buried in cemeteries along the western front, mainly in northern France and Belgium, while momuments were raised along the length and breadth of the United Kingdom in remembrance of each community's human losses. Importantly, both the cemeteries and the monuments represent key symbols of the loyalty of thousands of individuals to the British nation. The Arlington cemetery in the United States and the Arc de Triomphe in Paris are other clear examples of this memorialisation of the dead of the nation. Perhaps most powerful, in this respect, are the tombs of unknown soldiers that exist within most states. Benedict Anderson (1983: 9) has argued that 'no more arresting emblems of the modern culture of nationalism exist than cenotaphs and tombs of Unkown Soldiers'. Because of the anonymity of the soldier that lies under the tomb, these tombs can come to represent the more general sacrifices that all individuals (should) make for their nation.

Memorials to the dead of the nation are one powerful type of place that helps to focus the nationalist sentiment of members of the nation, but others also exist. Parliament buildings and monuments (see Plate 5.2), for instance, embody the citizenship of all individuals within a state, along with their membership of a nation. Other examples include national museums. These are seen as illustrations of the nation's historical development and are therefore a key method by which the nation can demonstrate its achievements to visitors, whether they are members of the same nation or others. The role of museums as generic places that play a key role in symbolising nations is elaborated upon in Box 5.1.

As well as certain generic places that are important to all nations, we also need to consider the specific places that possess a significant meaning for particular nations. What is important here are the particularities of the histories and geographies of a given nation, ones which give meaning and value to specific places. In the remainder of this section, we discuss two brief examples of places that play a significant role in symbolising or inspiring particular nations.

Our first example is the Cathedral of Christ the Saviour in Moscow. Dmitri Sidorov (2000) has examined the significance of this particular place for the Russian nation. Originally designed as a memorial to the great Russian victory against Napoleon's French

Plate 5.2 A statue commemorating Garibaldi, 'father' of the Italian nation
Courtesy of Rhys Jones

BOX 5.1 FOLK MUSEUMS AND THE MEMORIES OF THE NATION

National museums are key sites for any nation because they enable a nation to represent itself to its members and to the world. Particularly important are the folk museums that exist within many countries. Seen as the repositories of the folk culture of the nation, they are viewed as key means of representing the essential cultural truths about a particular nation (see Williams and Smith 1983). A good example of a folk museum is the Skansen Open Air Museum, situated in Stockholm, the capital of Sweden (M. Crang 1999). Skansen, the world's first open-air museum, was opened in 1891 as a way of preserving elements of the Swedish folk culture. Crang has noted how Swedish intellectuals at the time displayed a great deal of interest in the cultural past of the Swedish folk. These folk (and mostly rural) cultures were perceived as the only links between a Swedish nation and its past, especially since the cultural make-up of the vast majority of the Swedish population was being transformed as a result of the processes of modernity. Skansen today comprises a number of costumed workers who 'inhabit' the houses and farms drawn from all parts of the country. Importantly for Crang (1999: 451) the folk culture preserved in Skansen is set up as 'a timeless, interior Other within the modern nation'. Skansen, and other folk museums, therefore, illustrate the importance of preserving a folk past for the modern political and cultural communities of the nation. They help to emphasise how contemporary nations are allegedly connected, through ties of culture, with the earliest incarnations of the nation in the dim and distant past.

Key reading: M. Crang (1999).

forces in 1812, it was viewed as a way of memorialising the 'unprecedented zeal, loyalty to and love of the faith and Fatherland' (quoted in Sidorov 2000: 557). Obviously, therefore, from the very beginning the cathedral was perceived as a means of symbolising the commitment of the Russian people to their nation. What is equally crucial, according to Sidorov, is that the cathedral, since its construction, has reflected and symbolised broader political and national changes that have occurred in Russian/Soviet society. So, for instance, the original European design was quashed in the late 1820s because it did not tally with the new Tsar's desire for the cathedral to reflect national Russian architectural forms. Similarly, its demolition in 1931, as a result of the Bolshevik revolution, was carried out as preparation for the use of the location where it stood as a site for the construction of the secular Palace of the Soviets. In other words, by replacing the cathedral with the Palace of the Soviets, this one place in Moscow could be seen to represent the far broader changes occurring in Russian/Soviet nationalism as it shifted from one which emphasised a strong religious form of nationalism to one which was wholly secular in nature.

As it turned out, the Palace of the Soviets was never constructed. After the accession of Mikhail Gorbachev in 1985, discussions began concerning the construction of a new cathedral. Sidorov argues that this process, once again, has reflected broader political and national currents in Russia. Crucially, therefore, this one place in the centre of Moscow can be viewed as a key symbol of Russian nationalism. Throughout its period as a construction project it has illustrated some of the significant changes to have affected Russian society. Moreover, it has been viewed as a place that should help to symbolise and inspire the Russian/Soviet nation.

Our second example is from the United Kingdom. National identity in the United Kingdom during the 1990s, as in a number of other countries, has been challenged by the processes of globalisation and region-alisation. Importantly, it has been argued by British politicians that the British nation must respond to these challenges in a positive manner. Tony Blair, as Prime Minister, for instance, has argued that British national identity should 'not . . . retreat into the past or cling to the status quo, but . . . [should] . . . rediscover from first principles what it is that makes us British and to develop that identity in a way in tune with the modern world' (Blair 2000: 1–2). Part of this effort to reconstruct a new form of national identity for contemporary Britain can be seen in the context of the construction of the Millennium Dome. This was opened in 1999 as a means of 'illustrating the importance of past achievements and future challenges to national identity' (Taylor and Flint 2000: 222). Viewed as a means of representing the 'best of Britain' and 'all that was good about "Cool Britannia"', the Dome has played a key role in the efforts of the contemporary British nation to rebrand itself. Importantly, this central place in the national architecture of Britain was seen as an important cog in a far grander national scheme of projects and constructions that would help to celebrate a vibrant new British national identity. Admittedly, the Millennium Dome and its associated paraphernalia were viewed by much of the British public as a failure. None the less, what it important here is that this one place in the heart of London was viewed as one which was of crucial significance to the development of the whole concept of a new form of British nationalism.

Place – whether thought of in generic or specific terms – is, therefore, critical to any understanding of nations and nationalism. Places help to symbolise and sometimes inspire the nation. In many ways, certain places become important elements of the national imagination. Another geographical concept that plays a significant role in national imaginations is the landscape.

Landscapes of the nation

Along with place, the most potent way of imagining the nation is through reference to particular landscapes. By landscape we mean not just the physical environment but also the meaning and values that are ascribed to it by individuals or communities. Nations tend to view particular types of landscape as ones that represent the values or the essence of the nation. In this section we focus on these landscapes at both a conceptual and a more empirical level.

Generally speaking, nations tend to portray rural landscapes as ones that symbolise the nature of the nation. We can relate this to some of the themes raised by Colin Willams and Anthony Smith (1983) in their classification of the various dimensions of state territory. Two, in particular, emphasise the important of rural landscapes to the nation. By referring to the habitat of the nation, Williams and Smith draw our attention to the stress placed by nations on the belief that members come closer to the nationalist ideal if they live 'on the soil'. What this means is that individuals are more likely to adhere to nationalist principles if they live in rural, rather than urban, areas. Folk culture is also important to many nations: the rhythms of life in rural areas lead to the formation of peasant lifestyles and these are prized as manifestations of the true character of the nation. In both of these contexts, therefore, rural lifestyles and rural landscapes are to be cherished by the nation.

We see examples of the promotion of these two elements – folk culture and habitat – in the politics of the Welsh Nationalist Party, Plaid Cymru, in the inter-war period. Pyrs Gruffudd (1994: 69–70) has demonstrated how the party argued that Welsh people would have to 'return to the land' if they were to gain their rightful place as a moral nation. Significantly, part of this political strategy was based on a belief that the Welsh nation needed to live in rural areas so that it could avoid 'anglicised metropolitan values'. It was also based on the belief that it was only in rural areas that the Welsh *gwerin* (folk), or in other words the upholders of true Welsh national and moral values, lived. In many ways, Gruffudd's research echoes the work carried out on the national imagination of the west of Ireland within Irish nationalism (see Johnson 1997). Here, once again, for much of the twentieth century the west of Ireland was perceived as the main bastion of Irish national identity. This national imagination drew on linguistic geographies. The west was, by far, the least anglicised and most Gaelic-speaking region of the island. It was also based upon the religious differences that were thought to exist between a Catholic west and a Protestant north. Equally important was a romanticised understanding of the rural lifestyles that existed there among the Irish 'folk'. Though these ideas have been thoroughly deconstructed by Irish commentators, they represent a set of discourses that are still of great relevance to popular understandings of Irish nationalism.

As well as helping symbolise the purity of the nation, certain landscapes can be used as a means of emphasising the differences that exist between one nation and another. Rob Shields (1991: 182–99), for instance, has focused on the importance of the Canadian north for the constitution of Canadian nationalism. In one context, the Canadian north can be viewed in much the same way as Gruffudd and Johnson respectively view rural Wales and Ireland. For Shields, 'the "True North" is a common reference "point" marking an invisible national community of the initiated' (1991: 198). For instance, the Canadian historian W.H. Morton notes that an understanding of the Canadian north is imperative if people are to fully comprehend the nature of the Canadian identity (quoted in Shields 1991: 182). The rugged northern landscape – divorced from a southern, urbanised Canada – is seen to represent the essence of the Canadian nation. Shields argues, however, that the Canadian north is more than merely a symbol of the purity of the Canadian nation, for it helps to distinguish the Canadian topography and nation from that of the United States. For example, much of the nineteenth-century literature that described the Canadian north as a significant factor in the formation of the Canadian nation was consumed by audiences in the United States. In effect, the landscape of the Canadian north came to be used as a symbol of the necessary national difference that existed between the Canadian and US nations. The existence of this massive arctic hinterland in many ways enabled the Canadian nation to identify with other northern nations and states, such as Norway, rather than with the United States (Shields 1991: 198). In this example, landscape, as well as being a symbol of the nation, was a signifier of the differences between neighbouring nations.

In this section we have discussed the importance of certain landscapes for nations. They help to symbolise the essence of the nation in myth, literature and song. In the following section we discuss the equally important link between nation and territory.

Nation and territory: the homeland of the people

Nations, as is shown in Anthony Smith's definition, must 'share an historic territory'. Nations are, in effect, rooted in particular territories. Indeed, James Anderson (1988: 24) has eloquently argued that:

> the nation's unique history is embodied in the nation's unique piece of territory – its 'homeland', the primeval land of its ancestors, older than any state, the same land which saw its greatest moments, perhaps its mythical origins. The time has passed but the space is still there.

In this quote we see part of the significance of territory for the nation. A nation's territory helps it to commune with its past and to emphasise the strong links that have always existed between it and the land in which it now resides. Similar to place and landscape, therefore, territory can offer significant ideological succour to nations. A particularly striking example of this process exists in the context of the Jewish nation. During its long time in exile, and since the formation of the Israeli state, the territory of Israel has furnished the Jewish nation with much ideological support (Azaryahu and Kellerman 1999; Hooson 1994). Struggles over land between Jews and Arabs in contemporary Israel, therefore, do not merely represent attempts to increase the amount of land under one's control in a physical sense. They also represent ideological struggles for the control of the symbolic body of the nation.

Territory plays other important roles for a nation. In the first place, it has been argued that territory is the conceptual link between the nation and the state in the form of the nation-state (Taylor and Flint 2000: 233). Of course, the notion of the nation-state is the ideal of nationalism, the perfect political scenario in which the boundary of a state matches in an exact manner the geography of the nation. It is in these political contexts that both the nation and the state may feed off each other. The nation helps to legitimise the whole existence of the state, binding its citizens into an unswerving loyalty towards it. At one and the same time, the state exists to protect the members of the nation, ensuring that their national rights are promoted at the expense of the rights of the members of all other nations. In this way, territory can be viewed as the basis for the ideological and organisational marriage between the nation and the state.

Of course, boundaries are crucial elements in the constitution of territories, and this fact draws our attention to a second important link between nations and territories. Daniele Conversi (1995) has argued that the boundaries between nations are critical elements in the constitution of the nations that exist on either side of the boundary. For Conversi, nations are defined, at least in part, through a process of 'othering', in which the faults of neighbouring nations – whether real or perceived – are emphasised as a means of promoting the strengths and qualities of the 'home' nation. All nations engage in this act of 'othering' to a greater or lesser degree. The United States, in recent years, has attempted to contrast itself with the Japanese nation, most clearly with regard to economic and cultural practices. Nations from the neighbouring countries of Greece and Turkey have also symbolised each other's

nations in negative ways. This situation has been further exacerbated by the conflict between the two countries over the national status of the island of Cyprus (see Box 5.2).

Of course, in the contemporary world of cultural globalisation, nations need not define themselves against other nations, but rather against the perceived growing importance of a supranational sense of group identity. In many ways, this mirrors David Harvey's (1989a) arguments regarding the tendency for contemporary groupings to try to preserve their own individuality and distinctiveness as communities of people. Michael Billig (1995: 99) has noted, for instance, how much of contemporary British nationalism is couched in terms of a need to preserve a distinct British identity in the face of an ever-increasing 'Europeanisation' of identity within Europe. For instance, John Major, as leader of the Conservative Party, announced in 1992 that he would 'never let Britain's distinctive identity be lost'. His attempt here was to defend a proud and valuable national identity against what he and many others saw as the detrimental

BOX 5.2 GREEK AND TURKISH NATIONALISM

The history of the struggle between Greece and Turkey over the island of Cyprus has its roots in two main processes. First, the domination of the eastern Mediterranean by the Ottoman Empire for much of the modern period and the related repression of the Greek population. Second, the efforts of the British Empire to control the eastern Mediterranean as a bulwark against the growing power of Germany and Russia. With the collapse of the Ottoman Empire after 1918, there has been much conflict over the status of the island. In the period between 1918 and 1939 many Greek nationalists on Cyprus and the Greek mainland sought to gain independence for the island from British rule. The Turkish minority on the island were naturally wary of this prospect. The relationship between Greece and Turkey deteriorated further in the period subsequent to 1945, with much of the nationalist angst deriving from conflict over the status of Cyprus. Atrocities were committed on Greeks living in Turkey and numerous riots took place on the island itself. The creation of an independent Cyprus in 1960 did not solve the issue and, indeed, four years later all-out war between Greece and Turkey was only narrowly avoided. Today, it is only a large UN presence that prevents the escalation of hostilities between Greece and Turkey. During the whole of this period, there has been a steady deterioration in the relationship between the Greek and Turkish governments and peoples. To a large extent, both nations exist in opposition to one other.

Key reading: Calvocoressi (1991).

cultural and political influences emanating from European politicians and bureaucrats.

Here is another example, then, of the key role played by the nation's boundaries and territory in helping to shape the nature of the nation. Admittedly, Benedict Anderson (1983) has criticised this interpretation of nations as being too exclusionary, negative and regressive in nature. For Anderson, the fact that certain individuals may become naturalised members of other nations demonstrates that the boundaries of nations are not always fixed. In other words, the boundaries of nations may be viewed as places where national cultures and languages mix, rather than being places of exclusion and mutual denigration. Of course, such arguments resonate with recent debates within human geography regarding the necessity to view place as open to outside influences, rather than being closed to them (see Massey 1994). Laudable though these sentiments are, plainly, nationalist discontent and war are still grounded, in many cases, in the defamation and hatred of neighbouring nations.

In the three preceding sections we have discussed the key role played by the concepts of place, landscape and territory in ideologies of nationalism. As a result of this discussion we argue that geography and spatial themes are at the heart of any understanding of the nation. We further reinforce this claim in the final section of the chapter, where we focus on processes that contest the nation.

Contesting the nation

The impression given in definitions of nationalism, and also in various nationalist ideologies, is that the nation is a coherent and stable community of people, to which its members demonstrate unswerving loyalty. Within the ideology of nationalism, therefore, the nation is conceived of as a homogeneous group of people who have been indoctrinated with the nation's ideals. The physical boundaries of the nation demarcate the territorial extent of a group of people totally unified in their love of the nation. All members of the nation contribute to this notion of a unified 'imagined community' of people (Anderson 1983). Of course, in our

more lucid moments, we know that claims such as these are unlikely to mirror the reality of social existence, and that other spatial, scalar and social processes help to contest the ideology of nationalism. In this section, we want to discuss three ways in which the nation is contested. We do this through reference to: the way in which other aspects of identity fracture the nation; regional movements that operate within the boundaries of particular nations; and the concept of the local production of national identity.

At a very basic level all nations are intricately subdivided. This arises from the fact that every individual who exists as a member of a nation is also an individual who possesses a certain gender, religion, race, sexuality and so on. This means that there is the potential for them to engage with the nation in slightly different ways, and therefore to contest any unified vision of the nation contained within nationalist ideology. Some of these themes come to the fore in the work of Sarah Radcliffe (1999) on feelings of national identity within a lower-middle-class neighbourhood of Quito, the capital of Ecuador. She found that women and men experienced 'different trajectories of affiliation' to the Ecuadorian nation (1999: 217). This meant, for instance, that women were more likely to express their sense of national identity through reference to ideas of love of nation, while men spoke in more neutral terms of a sense of obligation to the nation. Furthermore, there were significant differences in the way in which male and female Ecuadorians engaged with the *mestizo* racial category. Originally, this was the name of a mixed racial group of European and indigenous people, but especially in the post-war era it came to symbolise 'an engagement in the urban, market-led and modernizing national society and an avenue for social advancement' (1999: 215). Importantly for Radcliffe, men and women in Ecuador engaged with this *mestizo* category in a different way from each other. On the whole, a lower proportion of women identified with the *mestizo* category, and those who did, did so in an ambivalent manner. Indeed, many women viewed themselves as white rather than *mestizo*, and identified with 'white' forms of language use, dress code and literature. For Radcliffe, therefore, women and men in Ecuador became entangled in Ecuadorian nation identity in different ways, as a result of their different engagements with issues of gender and race. This is one clear instance of the way in which individual identity can contest a homogeneous vision of nationalism.

A second way in which nations' attempts to promote themselves as unified bodies of people is shown to be a political and cultural fallacy is in the context of regional movements. Regional movements – operating at a spatial scale smaller than the nation-state – seek to increase the political, cultural or economic recognition afforded to them by a particular state or nation (see Paasi 1996). In some circumstances, regional movements may evolve into national movements and may seek independent status from the nation-state in which they reside. Regional movements' significance is that they demonstrate that nations are not coherent and unified communities of people. Nations' cultural and political dominance is sometimes challenged by other communities of people who feel themselves to be somehow different from the norm that is portrayed by the nation. Indeed, the fact that regional movements exist in a large number of nation-states in the contemporary world shows us that the common cultural community of the nation is in fact a myth: regional movements are seeking to subdivide and challenge the autonomy of nations in all parts of the world. There are few better examples of regional movements in the contemporary world than the Northern League, or the Lega Nord, in contemporary Italy. The significance of this regional movement is discussed in Box 5.3.

BOX 5.3 THE LEGA NORD IN CONTEMPORARY ITALY

The roots of the Northern League (Lega Nord) lie in the Lombard League and other local political leagues in northern Italy. The main reason for their creation was the desire to defend the perceived cultural and ethnic differences of northern communities from the cultural imperialism of the south of Italy. These various local leagues combined in 1991 to form the Lega Nord. From approximately this period onwards, economics rather than culture or ethnicity became the main guiding principle of the league. The northern region of Italy, containing the industrial cities of Milan and Turin, was portrayed as one that was being held back by an overbearing national government and a parasitic south of Italy. As Agnew (1995: 166) puts it, 'the message was *stiamo bene*, "we are fine as we are", without outsiders who undermine what "we" have achieved. Whether the outsiders were Africans or Sicilians almost seemed a matter of indifference.' The emphasis during this period, therefore, was on the creation of a federal system in Italy where the north would be able to thrive. Until this period the Northern League, with its emphasis on ethno-regionalism and federalism, illustrates the fundamental schism that existed within the Italian nation. There is a further twist to the story. In the period 1992–4 the Northern League, as a result of a political coalition, entered the Italian government and was immediately tranformed from a regional party into a national party with policies designed to shape the future of the whole Italian nation-state. In other words, the regional movement had become a national party. In this rare example, a regional movement gained enough political power to be in a position to alter the policies of the state within which it was contained. It also began to contribute to the reproduction of a slightly different view of Italian nationalism, one which was based on ideas of regionalism, federalism and privatisation. This example clearly shows the ways in which regional movements can contest the nation. It also illustrates the mutability of nations, since their nationalist ideologies can be appropriated, at times, by certain bodies operating within them.

Key readings: Agnew (1995, 2002b), Giordano (2001) and Putnam (1993).

The discussion above about the complex nature of identity and about regional movements begins to draw our attention to the role that geographical scale plays in the contestation of the nation. One way in which we can combine these two elements is to think about the local production of national identity. Michael Billig's (1995) work is especially useful in this context. He has urged social scientists to consider the banality of nationalism, or in other words, the way in which it is subtly reinforced on a daily basis as a result of small-scale, mundane and banal processes. National newspapers, for instance, repetitively use the first person plural pronoun as a means of referring to the people of the nation – 'us', 'we', 'our' and so on. The fact that these words are used daily, according to Billig, helps to reinforce the idea that a common community of people exists within a particular country. As such, the daily and mundane use of words such as these can be viewed as instrumental processes in the reproduction of nationalisms within a given country. A good example of the banality of nationalism – at a national scale – is the research that has focused on the symbols and imagery used on national currencies (see Gilbert and Helleiner 1999; Unwin and Hewitt 2001). Plate 5.3 illustrates another instance of banal nationalism, namely the inclusion of the words 'Je me souviens' or 'I remember' on all car registration plates in Quebec. Presumably these words allude to the historic struggle for the promotion of a Quebecois nationalism.

Plate 5.3 The banality of nationalism: remembering the sacrifices made for Quebec

Courtesy of Rhys Jones

The significance of Billig's ideas, therefore, is that they help to emphasise the mundane and everyday processes that help to mould individuals as part of the nation. We argue that many of these mundane and banal processes and happenings take place in local or small-scale settings. Research by Fevre *et al.* (1999) in Wales, for instance, has demonstrated how such process of reproduction could take place at small scales. For example, for many people in north Wales, it is the processes that operate within the local housing market, in which Welsh-speakers cannot afford to compete with English newcomers, that helps to engender within them a sense of Welsh nationalism. Similarly, it is individuals' experiences in pubs and bars – where people with varying linguistic abilities meet each other, and where arguments may take place – that enables them to shape their own interpretation of their national identity. As individual members of nations, we can all recall experiences from school, university, on the street or when socialising, that have made us think about our role and place in our nation in particular ways. For Rhys Jones, for instance, his experiences on the streets of Mold in north Wales forced him to re-evaluate his place within the Welsh nation. We argue that the logical outcome of these local reproductions of nationalism is the creation of nationalisms that vary slightly from place to place (see Edensor 1997; Jones and Desforges 2003). If we think place to be important in the shaping of our identity, then surely it will mean something slightly different to be a member of the US nation in Birmingham, Alabama, as opposed to Seattle, Washington. Or to be a member of the Brazilian nation in Rio de Janeiro, as opposed to a village in Amazonia. If this is the case, then we argue, once again, that nations are contested from within, for the simple reason that they incorporate numerous different places, and therefore numerous different types of people with numerous types of national identity within their boundaries.

The fact that nations can be reproduced at scales other than the national raises important issues regarding the impact of cultural globalisation on nationalism. The growing production, circulation and consumption of cultural messages at a transnational or global scale does not necessarily mean that nationalisms are being undermined within specific countries. Potentially, nationalism can be reproduced as a result of processes operating at any spatial scale, and therefore it would be unwise to predict the end of nationalism as a structuring principle for cultural communities. Nationalism is here to stay, at least for the foreseeable future, and as long as it still exists it will draw on geographical concepts of place, landscape and territory for its sustenance.

Further reading

The key starting point for geographical studies of the nation and nationalism is Williams and Smith, 'The national construction of social space', *Progress in Human Geography*, 7 (1983), 502–18. This paper, which in many ways represents the beginnings of geographical studies of the nation, elaborates on different geographical themes that are intimately related to nationalism.

Numerous studies have examined the significance of place, landscape and territory for the nation. For a good discussion and case study of the significance of particular places for the nation see Sidorov, 'National monumental-ization and the politics of scale: the resurrections of the Cathedral of Christ the Savior in Moscow', *Annals of the Association of American Geographers*, 90 (2000), 548–72.

Whelan's discussion of the politics of monuments in Dublin, although couched in terms of colonialism, also illustrates the importance of monuments for cultural identity; see 'The construction and destruction of a colonial landscape: monuments to British monarchs in Dublin before and after independence', *Journal of Historical Geography*, 28 (2002), 508–33.

The significance of particular landscapes for the nation has also been the source of much debate within geography. A good discussion of these themes can be found in Daniels, *Fields of Vision* (1993). This book gives an account of the use of rural images as a way of inspiring members of a nation, particular during times of tension and conflict, for instance wars.

More recently a number of authors have started to demonstrate the way in which nations are contested from within. A good example can be found in Radcliffe, 'Embodying national identities: mestizo men and white women in Ecuadorian racial-national imaginaries', *Transactions of the Institute of British Geographers*, 24 (1999), 213–26. See also Jones and Desforges, 'Localities and the reproduction of Welsh nationalism', *Political Geography*, 22 (2003), 271–92, for a discussion of how place-based identities can undermine notions of a coherent and uniform community of people within the nation.

6

Politics, power and place

Introduction

The nation is just one of the many different scales at which we define our identity and become engaged in political action. Below the level of the nation there are regions, states, provinces, counties, cities, towns, municipalities, parishes, neighbourhoods – many different *places* in which and through which politics occurs. This chapter examines how place provides a context for the formation of political identities and the identification of political interests, how political activity can be organised and mobilised around place, and how power within place is structured and exercised. In doing so, however, we seek to avoid two common traps that have sometimes ensnared political geographers in the past. First, we must be careful not to overstate the causal significance of place, but instead recognise that every place is constituted through wider social, economic and political processes. Second, in choosing to think about 'place' at a local scale – that is, to think about places as *localities* (see Box 6.1) – we must also recognise that 'the local scale' cannot be taken as a given entity but is socially constructed and that as such there is a politics of scale. These two arguments are discussed in more detail below before the chapter moves on to look at aspects of place-based community politics and power.

Why place still matters

It is a cliché of the western movie – the horseback stranger rides in off the dusty plain and enquires, 'Who runs this town?' Today many people, including many

---

BOX 6.1 LOCALITY

Despite its popularity within geography and sociology during the 1980s and early 1990s, there is no general consensus about the definition of the term 'locality'. For simple geographical delimitation researchers on the locality studies of this period tended to take local labour markets or travel-to-work areas as a 'locality' – but this really reflected the economic focus of the research. If political processes had been at the forefront, local political units might just as legitimately have been employed. This uncritical use of pre-defined areas opened the studies to charges of the unwarranted privileging of space (or 'spatial fetishism'), and Massey (1991: 277) argued that 'localities are not simply spatial areas you can easily draw a line around. They will be defined in terms of the sets of social relations or processes in question'. More controversial have been attempts to define locality as a concept – with disagreement about the attribution of causal powers to localities forming the core of the locality debates discussed below (see Box 6.2).

Key reading: Duncan (1989).

political researchers, would suggest that the correct answer should be 'Nobody around here.' Power, they would contend, can be attributed only to the state, or the media, or global corporations, or the invisible hand of capitalism. Place, to them, is unimportant in thinking about politics. This is the attitude of much political science research, which concerns itself with social classes and lifestyle groups but tends to conduct its analysis at a national scale with little attention paid to the effects of geography or spatial variation.

The marginalisation of place within political research stems from two separate critiques. The first is theoretical and has been largely associated with Marxist political economy approaches. Marxist theory emphasises structure over human agency and binds together social, economic and political processes and outcomes as part of an overarching capitalist system in which the need to reproduce and accumulate capital is the primary driving force. As such, the Marxist approach to urban politics conceptualised the city as a geographical entity produced and reproduced through capitalism, not as a neutral vessel in which autonomous local politics took place (Harvey 1973). Questions about the organisation, motivation and 'power' of urban elites and managers were sidelined as distractions from the structures through which the interests of capital were advanced. Marxist commentators reconceptualised the city as a site of capitalist oppression, where the agents of capital acted to produce favourable conditions for capital accumulation (see for example, Cockburn's 1977 thesis on the role of the *local state*) but also as a site of conflict, which is both produced by and helps to sustain capitalism (Cox and Johnston 1982).

Local politics, according to this argument, are primarily concerned with the resourcing and management of public services, such as housing, health care, public transport, education and social services, which are central to the reproduction of labour (Castells 1977, 1978). Conflict over these issues helps to displace class conflict from the work force to the city and distracts political opposition away from the global forces of capital that are the real sources of power (Castells 1983; Cockburn 1977). In the further development of this argument by Castells (1977, 1983), 'urban' is defined

by reference to these processes of labour reproduction rather than by any spatial characteristics, such that, contrary to the theories of Marxist geographers like Harvey and Cox, 'space' and 'place' were removed from the analysis of urban politics (Dunleavy 1980). However, this reduction of the urban to the social relations that define it was criticised, most notably by Urry (1981), who argued that the arrangement of social objects in space (such as the mix of classes in a particular town) can have an effect on wider social relations.

The second critique is based on empirical observation of the centralising and homogenising tendencies of late modernity. Local distinctness and autonomy, it is argued, have been eroded by a number of parallel processes. Globalisation has meant that localities are increasingly subject to the impacts of social, economic and political processes that operate at a higher level far beyond their control or influence; the greater mobility of people, advanced communications and cultural homogenisation have eroded local political cultures and mean that the same issues tend to frame politics in all localities; and in countries such as Britain the autonomy and power of local state institutions have been clipped by the centralisation of the state. However, beguiling as this argument may seem, research has shown globalisation and its attendant processes to be more complicated than this model would suggest. The impacts of globalisation are influenced and mediated by local factors, and globalisation can be accompanied by a simultaneous rescaling of power down from the level of the nation state to the local arena – labelled 'glocalisation' by Swyngedouw (1997) (see also Chapters 3–4).

Following these critiques, the question of how far the 'local state' – that is, the assemblages of state and governmental institutions that are organised at a local scale – is autonomous of the central state or how far it is an agent of the central state preoccupied many geographers and political analysts during the 1980s with mixed results. As Miller (1994) observed, 'those who attributed much autonomy to the local state tended to place more emphasis on the political process of decision-making . . . those who saw the local state as a functional arm of the central state focused more on economic relations' (Boudreau 2003: 181–2).

Included in these analyses were the 'locality debates' between British social scientists, which were linked with a programme of research that explored local responses to economic restructuring (see Box 6.2). The locality debates helped to re-focus attention on place within political and economic geography, but they eventually became too entangled in detail to provide any real conceptual pointers as to how the political geographies of localities should be researched.

More useful is the reworking of the concept of place by Doreen Massey in her article 'A global sense of place' (originally published in the magazine *Marxism Today* in 1991 and reprinted in Massey 1994). Massey starts from an account of time–space compression and globalisation and the 'increasing uncertainty about what we mean by "places" and how we relate to them' (Massey 1994: 146). She argues that time–space compression has a particular 'power geometry' in which some people, some communities, some localities are empowered and enjoy the benefits of globalisation while others are disempowered and disadvantaged. This leads her back to the question of place:

> How, in the context of all these socially varied time–space changes do we think about 'places'? In an era when, it is argued, 'local communities' seem to be increasingly broken up, when you can go abroad and find the same shops, the same music as at home, or eat your favourite foreign-holiday food at the restaurant down the road – and when everyone has a different experience of all this – how then do we think about 'locality'?
>
> (Massey 1994: 151)

The answer, suggests Massey, is to adopt a 'global sense of place' that is constructed not around political or administrative boundaries, but through the connections that link one place with other places:

> In this interpretation, what gives a place its specificity is not some long internalized history but the fact that it is constructed out of a particular constellation of social relations, meeting and weaving together at a particular locus. If one moves in from the satellite towards the globe, holding all

those networks of social relations and movements and communications in one's head, then each 'place' can be seen as a particular, unique, point of their intersection. It is, indeed, a meeting place. Instead, then, of thinking of places as areas with boundaries around, they can be imagined as articulated moments in networks of social relations and understandings, but where a large proportion of those relations, experiences and understandings are constructed on a far larger scale than what we happen to define for that moment as the place itself, whether that be a street, or a region or even a continent.

> (Massey 1994: 154)

Place, then, can be thought of as the intersection of a unique mixture of social, economic and cultural relations, some of which are local in character, some of which have a global reach. As such, place does therefore matter in political analysis because the distinctive ways in which these intersections are constituted, and the ways in which different actors engage with the particular combination of relations in a particular place, have real political effects. However, as the last quote from Massey indicates, a 'place' need not exist at a specific scale, but may be anything from a street to a continent. Hence if we want to think about places in terms of localities, that is, places that are defined as existing at a 'local' scale, we need to understand not just what is meant by 'place' but also what is meant by 'local'.

The political construction of scale

Traditionally, an examination of sub-national politics in a political geography textbook would include a discussion of the governmental units that existed at the different geographical scales – local, regional and national (e.g. Prescott 1972; Glassner 1996). Similarly, researchers studying local politics would often take the territory of a local government institution as their unit of analysis (e.g. Cooke 1989; Dahl 1961). In both cases the existence of a 'local scale' is taken as a given and is considered to be unproblematic. Little attention is paid to questions about how the scale was fixed, how

BOX 6.2 THE LOCALITY DEBATES

The locality debates are associated with a research initiative in the 1980s that explored local responses to economic restructuring in Britain. As the programme's director, Phil Cooke, explained, it sought to address the 'difficult question' about late twentieth-century life: 'While people's lives continue to be mainly circumscribed by the localities in which they live and work, can they exert an influence on the fate of those places given that so much of their destiny is increasingly controlled by global political and economic forces?' (Cooke 1989: 1).

In fact there are two questions hidden within this conundrum (see Duncan and Savage 1991). First, *what importance can be given to locality in conducting research on political and economic geography?* Duncan (1989) argued that localities matter because there are locally contingent factors that alter the nature of social structures in particular places and consequently influence the direction of social and political action. This assertion was supported by empirical evidence from the localities research that showed, for example, that in the north-east of England the strength of organised labour in industries such as shipbuilding and coal mining had built a left-wing political culture founded on trade unionism, which became undermined as these staple industries declined, muting the local political response to change; while in Birmingham the paternalism of major employers bred a more moderate local political culture, but also a tradition of work-based social associations such that it proved difficult to organise local responses to unemployment outside the factory (Cooke 1989). However, other geographers, such as Neil Smith (1987), expressed concern that the locality studies represented a covert return to empiricism and that locality research distracted from more 'worthy' global issues.

Second, *how does locality make a difference?* Answering this question proved more divisive among the localities researchers. Cooke, for example, held that localities are a form of social agent and that 'proactive' localities can have the power to cause socio-economic change (Cooke 1989). However, critics dismissed this notion as dangerously close to suggesting that geography determines social patterns, and argued instead that spatial variations can be incorporated into the analysis of social processes only as is appropriate for the research problem concerned. The importance of space may vary depending on the problem being investigated, and pre-defined territories such as the areas of local government institutions could not be uncritically assumed to be the appropriate spatial framework for research (Duncan 1989; Duncan and Savage 1989, 1991; Gregson 1987). A third approach adopted an intermediate position, acknowledging that pre-defined territories could provide the context for research on localities but rejecting any suggestion that such localities existed as 'social objects' with a capacity to determine social outcomes (Urry 1987; Warde 1989).

A related debate in the United States asked a similar question about how much agency could be attributed to localities and local institutions in shaping political and economic outcomes. Work on urban economic development had highlighted the importance of 'parochial capital', such as rentiers who are tied to a particular locality, in the coalitions or 'growth machines' that created favourable conditions for investment and development (Logan and Molotch 1987). The economic interests of 'growth machine' members are hence perceived to be intrinsically entwined with those of the locality, producing a system in which flows of capital are influenced by competition between localities for investment. Cox and Mair (1988) consequently advanced the concept of 'local dependence' as being central to understanding the local politics of US cities. Locally dependent firms form business coalitions to stimulate investment which are supported by local governments because they are themselves locally dependent. Cox and Mair recognised that the resulting developments may sometimes threaten the interests of local people and could potentially be opposed. Such conflict, they argue, is generally avoided by business coalitions appealing to the local dependence of communities and re-casting concepts of local community 'in a form that better suits their needs' (p. 317) by eliding community with locality and presenting the interests of the community as being threatened by the competitive advantage of other localities.

Key readings: The findings of the British localities research are summarised in Cooke (1989). For more on the localities debate in Britain see Duncan and Savage (1991) and other papers in the same issue. For more on the locality as agent thesis see Cox and Mair (1988, 1991).

the boundaries were drawn, how the governmental responsibilities and political issues appropriate to the local scale were decided, and how the local engages with other scales. Yet, as the discussion of 'place' above demonstrates, localities cannot be understood as neatly bounded administrative territories, and places are intrinsically multi-scalar, constituted by social relations that range from the parochial to the global.

All this points to the contingency of scale and to its social construction (see Box 6.3) through both lay and state practices. Marston (2000) proposes three central tenets that underpin this approach. First, as noted above, scale is not an existing, given entity awaiting discovery, but rather the differentiation between geographical scales 'establishes and is established through the geographical structures of social interactions' (Smith 1992: 73; see also Delaney and Leitner 1997). Second, the ways in which scale is constructed have tangible and material consequences. They are not just rhetorical practices but are inscribed in both everyday life and macro-level social structures. Third, the framings of scale are often contradictory and contested and are frequently ephemeral (see also McCann 2003). Indeed, the contestation of scale recognises that the fixing of scale is in itself a political act practised by both state and non-state actors. The state routinely constructs scales, as it creates and restructures local government institutions, as it formulates and implements policies, and as it decides which issues are appropriately dealt with at which scale (Brenner 2001). This is part of the state's spatial strategy that enables it to govern (Jones 1997; see also Chapter 3). But scales of political action are also constructed by non-state actors. Herod

(1995, 1997), for example, illustrates the engagement of trade unions in contesting the scale at which bargaining with employers is most advantageously fixed. Similarly, grass-roots activists contribute to the construction of the local scale through the way in which they mobilise discursive representations of their neighbourhood or community, and through the issues that they select as the focus of local activism. McCann (2003), for example, examines how grass-roots Latino activists contested the scaling of new neighbourhood planning areas by the city authority in Austin, Texas, because the territories of the suggested units did not fit in with their spatial imagination of their community. Martin (2002) similarly discusses the role of activists in defining a neighbourhood public sphere in St Paul, Minnesota, and the use of gender-essentialising discourses of safety and parenting to position household and family issues as community concerns. It is in this sense that we can talk about not just the scales of politics, but also the politics of scale.

Local politics, therefore, are in part concerned with the construction of the local scale, and the scale of local politics is in part defined by those who participate in it. But need local politics be tied (and restricted) to any fixed territorial area when scale is so contingent, and when place is constituted by dynamic mixes of wider social and economic processes? To answer this question, Cox (1998) makes a distinction between the spaces of dependence of local politics and the spaces of engagement of local politics. Spaces of dependence are 'defined by those more-or-less localized social relations upon which we depend for the realization of essential interests and for which there are no substitutes elsewhere;

BOX 6.3 SOCIAL CONSTRUCTION

The concept of social construction suggests that 'things' (either material objects or abstract entities) do not have a preordained, intrinsic, 'true' definition or meaning, but that things are ascribed with meaning by and through social interactions. In other words, the meanings of things are socially constructed. Our knowledge about things does not therefore involve the discovery of truth, but rather knowledge is constructed within the social context of the enquiry and is informed by the prevailing beliefs, practices and experiences of the time, place and people involved. As such, the meanings that we give to things are always contested, contingent and ephemeral.

Key readings: Braun and Castree (1998), especially the introduction, and Sismondo (1993).

they define place-specific conditions for our material well being and our sense of significance' (Cox 1998: 2). The space of dependence for local government is its territory or jurisdiction, but for other agents the space of dependence might be a labour market, or a local economy, or some other geographical unit. Hence, for any particular spatial point, 'local politics' will involve many different institutions, each with its own space of dependence fixed at different scales.

However, local actors – people, firms, state agencies, campaign groups, etc. – also have to engage with other 'centres of social power' that exist outside their space of dependence – local, regional and central government, transnational corporations, the national and international media, and so on. Cox defines these relationships as the 'space of engagement', which he further conceives of as a network, unevenly penetrating different scales and areas (see also Low 1997). Cox (1998) illustrates the construction of a scale of engagement with reference to a land use conflict in southern England, described by Murdoch and Marsden (1995). In this, opposition is organised to a proposal for a gravel extraction pit at a site near Chackmore, Buckinghamshire, identified as part of a national strategy and rationalised in terms of national mineral needs. As such, the 'space of engagement' was constructed by the state at the national scale. The opposition group therefore could not fight the proposal at the local scale, but needed to construct its own national-level network of agents. It did this by commissioning a hydrological study that showed that the development could lower

the water table, draining the ornamental ponds in the renowned parkland at nearby Stowe Park. This enabled the protest group 'to build a national-level network around an alternative representation of the site: not [so] much important to the national economy as to the national heritage' (Cox 1998: 8), enrolling elite national actors whose influence was sufficient to stop the development.

Places that exist as intersections of diverse social, economic and cultural processes, fixed at a socially constructed 'local scale', hence frame the practices and concerns of 'local' politics. Yet local politics are not bounded by this apparent spatiality. They are influenced by events that may be distant in space and time. The local politics of localities with large diasporic communities, for instance, are often informed by the politics of the homeland. In summer 2003 the town of Huntington on Long Island, New York, voted to remove the family crest of the English ruler Oliver Cromwell (who was born in the town's near namesake of Huntingdon, England) from its coat of arms after lobbying by Irish-American groups motivated by Cromwell's brutality in the seventeenth-century annexation of Ireland (Buncombe 2003). Similarly, global corporate politics have a particular significance to residents in single-industry towns. Moreover, the pursuit of local politics may on occasion require the construction of 'spaces of engagement' that enrol actors outside the locality, including at regional, national and international scales. Indeed, the ability of local communities to participate in multi-scalar spaces of engagement can be empowering and emancipating and can reflect back to shape local spheres of political activity. Perreault (2003), for example, describes the processes of political organising in the Quichua community of the Mondayacu region of Ecuador. Here the construction of a local scale of political organisation, including the creation of new local institutions and territories, is intrinsically connected with the participation of community-based indigenous groups with transnational organisations and agencies in multi-scalar networks that position the local activity as part of a wider development process. In this way, Perreault notes, mutli-scalar networks link local and trans-local processes, producing and consolidating social constructions of place and 'thickening' the emerging civil society.

Having positioned local politics, the remainder of this chapter proceeds to focus on two distinct aspects of the interaction of politics and place. First it considers how the characteristics of a place (and of the communities associated with that place) inform the political identities, interests and perceptions of residents. Second, it examines how power and leadership are structured within localities (or communities).

Place, community and political interests

Places are not the same as communities. A community is a group of individuals who are bound together by a common characteristic or a common interest and who enjoy a relatively high degree of mutual social interaction. Communities are defined by shared meanings and enacted through established and routine practices that occur within particular spaces and structures including 'both the material sites filled by communal activities, and the symbolic and metaphoric spaces in which people connect "in community" even while existing in different physical or social locations' (Liepins 2000: 28). Communities, therefore, are frequently identified with a particular place, but need not be. Many communities do not have a geographical identity (although all have a geography), but represent a social group with members in many different places.

Moreover, few localities are associated with a single, homogeneous community. Rather, there are commonly many different, overlapping, communities in a locality, such that local politics often revolves around conflicts or disputes between different communities.

It is, however, the characteristics of the dominant community (such as ethnicity, religion, class) as much as the characteristics of the place (the physical, economic and administrative structure) that constitute the politics of place (see Box 6.4). This happens in a number of ways. First, communities are a source of identity for their members and thus may be treated as a collective that may take political actions *en masse*. For example, a sense of collective identity may lead members of a community to vote for a co-member standing for higher office (the so-called 'friends and neighbours' effect, see Chapter 8). Second, communities are filters through which people view the wider world, judging political issues in terms of their impact on the community as a whole and identifying their own personal interests with the interests of the community. Third, as social collectives, communities have their own internal power structures, leaders and conflicts. Fourth, because communities rely on ideas of collective identity, they can also become exclusionary forces, promoting conformity and suppressing difference. As such, for example, communities that define themselves at least in part in terms of conservative religious identity or masculine culture may be intolerant of non-heterosexual behaviour (Joseph 2002).

The post-communist transition in Central and Eastern Europe provides an interesting example of how the local politics of communities both reflect and shape individuals' perceptions of wider change, not least because these processes occurred in parallel with the social construction of a new local scale of political action. Fiona Smith (2000) describes the case of a run-down housing estate in Leipzig, East Germany. Following the downfall of the communist regime, the strategy for the renewal of the estate was framed by the Western discourses of democracy and the market. The former was mobilised through community participation in the planning process. As a participatory local civil society did not exist under the

BOX 6.4 CULTURE AND LOCAL POLITICS: POPLAR IN THE 1920s

Poplar was formerly the easternmost borough in the East End of London. Its crowded and largely sub-standard housing was the product of rapid urbanisation in the late nineteenth century, its population predominantly employed in the nearby London docks and in large engineering and drink and food-processing factories. In the early 1930s it was identified as the poorest borough in London, with a quarter of the population living in poverty. Local politics were dominated by a radical left-wing local Labour Party which frequently took a maverick line, coming into conflict with the national Labour Party and earning Poplar a reputation as a 'little Moscow'. In one incident in 1921 Poplar refused to pay precepts to London-wide authorities, including the police force, spending the money instead on poor relief within the borough. By the end of the decade, however, the party had shifted rightwards and Poplar council became embroiled in corruption. Rose (1988) argued that Poplar's initial radicalism and later corruption could not be explained simply by its economic structure, but that the local culture also had to be considered. She explored the linkages between Poplar's politics and five aspects of its cultural life. First, the prevalence of home working and hired labour in Poplar's economy made trade unionisation difficult and meant that Poplar's *class politics* was mobilised not around trade unions but around a number of militant left-wing political groups which had a radicalising effect on the local Labour Party. Over time, however, councillors became more distanced from their working-class base, lessening their sense of class identity and their radicalism. Second, a strong sense of *neighbourhood identity* and the importance of the family had promoted community solidarity which informed the radical politics. Later, nepotism contributed to the growing corruption. Third, the *religious* fervour of local activists supported not only a radical Christian Socialism but also a moral politics which was often non-socialist and sometimes racist in its objectives. Fourth, over time *voting Labour* became a matter of habit for Poplar residents – part of their culture – such that they continued to support the party even as its radicalism waned and corruption grew.

Key reading: Rose (1988).

---

communist regime, community activists first had to construct the neighbourhood as a space of political activity:

> Neighbourhood activist groups worked to create local public spheres, holding events which, in using a range of local public spaces, such as halls, squares, parks and schools, displayed to the local population and to wider public the active construction of local agendas.
>
> (Smith 2000: 138)

The latter discourse, that of the free market, structured the way in which the redevelopment was implemented and managed, and proved more contentious for residents. Accepting the free market meant abandoning the principle of equal rents and allowing market forces

to dictate property prices, and, consequently, the pace and type of housing renewal. In the new system, the responsibility for renewal lay with the property owners, with the state able to facilitate action but not to prescribe or carry out the action itself. Residents used to the imposition of standardised work programmes often had difficulty adapting to the new system and expressed disappointment that promises of better housing had not been realised. These experiences at the local scale informed the residents' perceptions and opinions of the larger-scale changes involved in the transition from communism. Moreover, experiences of the transition differed between localities:

> The problems in this neighbourhood were not found uniformly across Leipzig. In each neighbourhood the nature of political contests and local problems

varied with differing combinations of factors such as housing stock conditions, socio-economic profile of the population, speed and dimensions of property restitution, time-frames for planning and local activism, the effects of capital (speculation, continued disinvestments, pressures for land-use change) and past and present forms of local activism.

(Smith 2000: 135)

Thus the construction of the neighbourhood as a political space that combined the physical characteristics of the locality (the housing stock, etc.) with the social characteristics and the agency of the community created both an arena in which citizen participation could be mobilised and a set of experiences through which the wider process of political and economic transition could be understood.

Local power and leadership

The sections above have advanced the argument that place is important in political analysis and cannot be ignored. There are place-specific characteristics that mediate the impact of wider social, economic and political processes on individual localities, and that influence the way in which people engage with these wider processes. Moreover, this indicates that there is a local scale of politics that enjoys a degree of autonomy and agency and which in turn raises questions about the way that power is produced, circulated and exercised within localities or communities. All communities (and thus, by extension, all localities) have 'leaders' who are responsible for decision making in the day-to-day government of the community, who organise the local political sphere and who represent local interests to external actors. There is always, therefore, some concentration of power within a community.

The analysis of community power, however, is more complex than may initially be apparent. It may seem straightforward to ask 'Who has power in this city?' Direct the question to people in the street and they will probably give you a straightforward answer – the mayor, or a major employer, or perhaps even some kind

of 'local mafia' or political elite. But think for a moment about these replies. Is it really the mayor who has power, or the people who elect her or him? Are they 'powerful' as an individual, or only when 'in character' as mayor? Or does their 'power' rest on there being other people – administrators, council workers – who will carry out actions as instructed? If a company is deemed to be powerful, do we mean the company itself or the directors or the shareholders? How does it actually exercise its 'power'? And what about the so-called elite? Who are they? How did they get there? What 'power' do they actually have over other citizens on a day-to-day basis? *Come to think of it, what do we mean by 'power' anyway?*

These questions formed the crux of the 'community power debate' in American social science in the 1950s and 1960s between the 'pluralists' and the 'elitists' (Box 6.5). The debate was partly conceptual – is power a property that can be possessed or does it exist only when it is exercised? – and partly methodological, the elitists starting with the individuals with a reputation for power and the pluralists starting with decisions that demonstrated the exercise of power. Both approaches revealed elements of the way in which local power and politics work, but neither convincingly revealed the whole picture. By the 1970s the community power debate was judged to have stagnated into stalemate (Harding 1995).

Since the 1980s, however, a more comprehensive approach to community power has been developed under the aegis of urban regime theory. As Lauria (1997: 1) describes, urban regime theory 'dispenses with the stalled debates between elite hegemony and pluralist interest group politics, between economic determinism and political machination and between external or structural determinants and local or social construction' by shifting attention from decision making to the setting and achievement of strategic goals. In essence, urban regime theory 'asks how and under what conditions do different types of governing coalitions emerge, consolidate, and become hegemonic or devolve and transform' (1997: 1–2). Its central thesis is that in order to maintain stable conditions for capital accumulation, local regimes are formed which draw together coalitions of institutions, interest groups and

BOX 6.5 THE COMMUNITY POWER DEBATE

The community power debate between elite theorists and pluralists was partly methodological, partly theoretical and partly interpretative. Methodologically the pluralists criticised the 'reputational approach' taken by Floyd Hunter in the most notable 'elite' study of community power in Atlanta, Georgia, which involved asking individuals in prominent positions in the city to rank other prominent individuals according to their reputation. Robert Dahl, the leading pluralist, argued that this approach predetermined the results of the study because by starting with the assumption that a power elite could be identified by reputation it did not allow for the possibility that there was no elite. Pluralists further criticised the reputational approach for suggesting that individuals held power while ignoring the power vested in jobs and social roles and for failing to present evidence of power actually being exercised (see Dahl 1958, 1961; Harding 1995; Polsby 1980).

These latter two criticisms reflect the theoretical differences between the elite theorists and pluralists over the concept of power. Hunter had employed the classic definition of power as a property that could be *possessed*. For Dahl, however, power needed to be *relational* – summarised in his own definition (see also Lukes 1974, 1986): 'A has power over B to the extent that he can get B to do something that B would not otherwise do' (Dahl 1957: 203). Thus Dahl's own methodology in New Haven was 'decisional' or 'positional' in approach. Concentrating on the three issues of urban redevelopment, public education and political nominations, his team traced back the germination of particular policies and outcomes to question why particular decisions were made and who participated in the decision making (Dahl 1961).

By starting from different theoretical standpoints, and by employing different methodologies, Hunter and Dahl were led to different interpretations of the urban political systems they studied. Hunter in Atlanta 'identified' a power structure controlled by a small and largely invisible policy-making elite comprised almost exclusively of key business leaders together with the mayor. Superficially, government was exercised through the city council and other public agencies but, Hunter contended, decision-making power rested with the business-led elite, not with elected officials. In contrast, Dahl argued that New Haven had a 'stratified pluralist' system. Only a small number of people were found to have direct influence in decision making and these belonged to a politicised stratum of the city's population. However, Dahl argued that this politicised stratum did not represent a single power elite, as there was no overlap between the individuals with influence in the different policy arenas that he studied. Moreover, while most citizens were not politically active, the pluralists contended that they possessed a 'moderate degree of influence' over their elected leaders whose decisions reflected 'the real or imagined preferences of constituents' (Dahl 1961: 164).

Dahl's pluralist model has also been criticised on both empirical and methodological grounds. Empirically it is flawed because not all interest groups are able to compete on equal grounds, not all voters vote and the identification of a set of competing oligarchies does not equate to a pluralist system (see, for example, Newton's (1969, 1976) work on Birmingham, England). More fundamentally, Bachrach and Baratz (1962) claimed that Dahl, like Hunter, had predetermined his findings by making particular assumptions about power when selecting his methodology. They disputed Dahl's definition of power, suggesting that the exercise of power did not need a decision to be made, but that power is also exercised through the shaping of social and political values and institutional practices to limit the scope of political action. Thus, as they attempted to demonstrate in a study of Baltimore (Bachrach and Baratz 1970), the 'power' of urban policy makers is restricted to issues and outcomes that are relatively innocuous to the interests of the really 'powerful'. This thesis proved to be highly controversial, was fiercely rejected by pluralists (see Wolfinger 1971), and created a conceptual deadlock in community power studies that persisted until the late 1980s.

Key readings: Hunter (1953) and Dahl (1961).

political leaders around the pursuit of particular goals. Such regimes are contingent in that they must respond and adapt to changing social, economic and political circumstances (both local and external) and therefore can evolve in their membership and strategy. As there are a number of different strategies that local regimes can adopt (for example, they can be entrepreneurial or anti-development) local factors can shape the form of regimes and the subsequent policy outcomes.

Community power in practice: a case study of Atlanta

Significantly, one of the most important empirical applications of regime theory was undertaken in Atlanta, the state capital of Georgia in the south-east United States and the site of Hunter's research three decades earlier. In Hunter's original study he identified

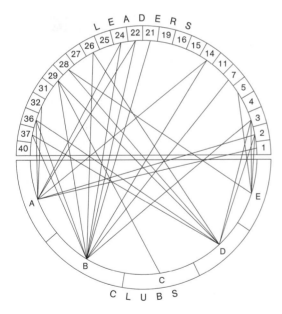

Figure 6.1 Interlocking club memberships held by members of Hunter's power elite in Atlanta

Source: Hunter (1953: fig. 7), copyright 1953 by the University of North Carolina Press, renewed 1981 by Floyd Hunter, reproduced by permission of the publisher

an elite of forty individuals across four sectors – government, business, civic associations and 'society' activities. The elite were fairly homogeneous in their social background and attitudes – they were mostly men, and all white. (Hunter identified a separate leadership structure for the black community.) Eleven were company directors, seven bankers or financiers, five lawyers, five industrial managers, four senior government personnel, two labour union leaders and one a dentist. The remaining five had sufficient private wealth to support full-time leadership of the city's civic and social organisations. They tended to live in the same desirable neighbourhoods in the north of the city, and were acquainted with each other through interlocking company directorships and club memberships (Figure 6.1). As Hunter (1953: 10) observes, their interactions formed particular geography:

> There are certain places in which they make decisions and formulate policies to meet the many changing conditions that confront them. In locating these men of power in a community one finds them, when not at home or at work, dividing their time between their clubs, the hotel luncheon and committee rooms, and other public and semi-public meeting places.

Not all of Hunter's elite were actively involved in policy making. Hunter identified a core group of 'power leaders' comprised exclusively of business leaders, together with the mayor, who were regularly consulted and involved in the development of policy. Hunter is less clear on precisely how this influence was exercised, but places the emphasis less on direct intervention than on the part played by the elite in maintaining consensus on policy issues and in anticipating challenges and changes in public attitudes.

Hunter returned to Atlanta to conduct a second study in the early 1970s (Hunter 1980). Employing a similar methodology as in his original study, he again identified an elite of forty leaders. Of these, seven had been on Hunter's list in 1950, including R.W. Woodruff of Coca-Cola. Several others were the children or relatives of names on the 1950 list, members of what Hunter labelled 'Atlanta's historically powerful families', the influence of some of which could be traced back to the city's foundation and development in the mid-nineteenth century. Hunter also identified connections between members of the two lists through thirty-six corporations and organisations, including Coca-Cola, Delta Airlines, banks and newspapers. Hunter hence concluded that the elite was still business-dominated and still operated through the interaction of social and professional networks, but he also went further in discussing how the elite's influence was exercised, linking elite members with specific policy achievements and recording the extensive representation of elite members, historic families and key corporations and organisations on the board of Central Atlanta Progress (CAP), the body responsible for redeveloping the downtown. Yet Hunter also found changes in the power structure. New leaders were emerging from new sectors, particularly property development; most significantly, the African-American community was no longer disengaged from the city elite. Leaders of the black community were now city leaders, and in 1973 Atlanta elected its first black mayor.

However, Hunter ultimately failed to explain these changes because he was unable to offer an explanation as to why the power structure existed in the form he described or why the elite had the members that it had. To explore this bigger picture, Stone (1988) structured his analysis of Atlanta politics around a new concept of power, 'pre-emptive power', or 'power as a capacity to occupy, hold, and make use of a strategic position' (p. 83). 'Pre-emptive' power is about *power to* do things, not about *power over* others. It is achieved through the blending together of resources in order to create a 'capacity to act', and as such is described as the power of *social production*, not *social control*. Thus, in Stone's (1988) reinterpretation of Hunter's study, the identification of a policy-making elite is significant not because its members can be said to possess power, but because the networking and interactions through which Hunter describes the elite as working represent the interaction of individuals with access to different sets of resources in order to form alliances with the capacity to act on specific policy issues. The regime approach goes further, proposing that pre-emptive

power is stabilised through the establishment of informal arrangements between governmental and other local organisations to constitute a *regime* 'with access to institutional resources that enable it to have a sustained role in making governing decisions' (Stone 1989: 4). Furthermore, within successful regimes there is a small core group of actors who are repeatedly involved in making key decisions and who form a 'governing coalition' that holds the regime together. Although the 'governing coalition' superficially sounds suspiciously like Hunter's elite, Stone distances it from the negative connotations associated with the term 'elite' by stressing that a governing coalition does not 'rule in command-and-control fashion' (Stone 1989: 5), but rather is concerned with the co-ordination and mobilisation of resources.

Hence Stone acknowledged the existence of Hunter's elite in the 1940s and 1950s, with informal interaction through organisations such as the Piedmont Driving Club, Capital City Club and Commerce Club providing a private operating space for a 'governing coalition' of business and political leaders. However, Stone contends that such elite interactions were not sufficient on their own and that the establishment in 1941 of the Central Atlanta Improvement Association (later CAP) was significant in creating a vehicle through which the big business interests of the elite could override the interests of small business by bypassing traditional business groups such as the chamber of commerce. Through elite networking and the CAP the governing coalition was able to hold together a stable regime that dominated Atlanta politics throughout the 1940s and 1950s. The elite had allies in the leadership of the black community, but the relationship was one of black subordination. The active members of the governing elite were as one leading figure described:

Almost all of us had been born and raised within a mile or two of each other in Atlanta. We had gone to the same churches, to the same golf courses, to the same summer camps. We had played within our group, married within our group, partied within our group, and worked within our group. . . . We were white, Anglo-Saxon, Protestant, Atlantian, business-oriented, non-political, moderate, well-bred, well-educated, pragmatic, and dedicated to the betterment of Atlanta as much as a Boy Scout troop is dedicated to fresh milk and clean air.

(Stone 1989: 56)

During the 1960s, however, a younger, more radical black leadership rose to prominence, increasing friction between the governing coalition and the black community. The governing coalition was losing strength and was vulnerable to occasional defeat, especially on issues where the black community could be mobilised. Coupled with national events and the growth of the black community as a proportion of the city population, this vulnerability led some in the business elite to realise that their own future influence rested on building alliances with the black leadership. In consequence, Atlanta endured a period of instability, as racial politics created divisions within the governing coalition between integrationists and segregationists, before the integrationists were able to forge a new governing coalition between white business leaders and black community leaders.

The 1970s brought a new challenge from the mobilisation of a neighbourhood movement that was able to threaten the fragile bi-racial coalition. However, as the neighbourhood movement was overwhelmingly white and frequently anti-development in its campaigns, it was unable to forge alliances with either the black community or the business elite and rather provided a rationale for their co-operation. Thus by the early 1980s the neighbourhood movement had faltered, and the governing coalition of white business leaders and black community leaders had become entrenched as the core of a stable urban regime. In many ways, Stone notes, this coalition was a revised version of the older 1950s coalition:

Blacks no longer occupy a subordinate position, but neither are they dominant. Nor is black participation in the coalition inclusive; just as before, the black middle class are the political insiders. However, that group has expanded to include not only public officeholders, but also blacks in white-owned corporations, such as Coca-Cola, Delta, and Rich's Department Store. The main difference

between the periods is that blacks control city hall, whereas earlier they only bargained with city hall as one of the city's voting blocks.

(Stone 1989: 135)

The consequence of this accommodation, Stone argues, was that neither partner had unfettered control of the policy agenda. Rather only policies that found a middle ground could be successful, while radical black community groups, jobs advocates, and white neighbourhood and preservationist groups all found themselves excluded from the policy process:

Atlanta's coalition between black middle class leaders and white business interests is no simple matter of giving the business elite what they want. Its chief policy thrust – a full-throttle development with almost no restrictions on investors, combined with strong encouragement and opportunities for minority businesses – brings the coalition together and promotes co-operation. Projects that meet these criteria move ahead despite enormous obstacles and numerous pitfalls; those that lack these ingredients make little headway.

(Stone 1989: 159)

Elite networks and discourses of power

Urban regime theory provides a useful framework for exploring the enrolment of particular groups into government coalitions and the selection of policy strategies, but it is less helpful in offering explanations about which individuals rise to prominence within regimes, and the extent to which they are representative of the wider community. This issue is significant because there is growing political awareness of the inequalities of representation in local governance. For example, only 22 per cent of US cities with a population of over 100,000 had female mayors in 1999. The two largest cities, New York and Los Angeles, have never elected a woman as mayor and women make up less than a third of city council members in New York, Los Angeles, Chicago and Boston (Table 6.1).

Table 6.1 Gender and local governance in the United States and Britain (%)

Local government members	n	Male	Female	Data n.a.
United States				
Mayors of cities of 100,000+ population, 1999	218	77	21	2
New York City Council members, 2002	51	76	24	
Los Angeles City Council members, 2002	15	67	33	
Chicago City Council aldermen, 2002	50	70	30	
Boston City Council members, 2002	13	85	15	
Britain				
County and district councillors, 1986	n.a.	81	19	
Council leaders, 1993	448	68	8	25
Council chief executives, 1993	448	97	3	
Greater London Assembly members, 2002	25	60	40	
Housing Action Trusts, board members, 1994	n.a.	54	46	
Training and Enterprise Councils, board members, 1994	n.a.	88	12	
District Health Authorities, board members, 1994	n.a.	62	38	
National Health Service trusts, board members, 1994	n.a.	60	40	

Sources: *World Almanac* 1999; city council Web sites; Byrne (2000); Tickell and Peck (1996)

In Britain, only 19 per cent of local councillors are women, and only 3 per cent are from non-white ethnic backgrounds (Byrne 2000). Similar patterns of skewed representation exist with respect to social class and age.

The gender bias of elected local government is exaggerated still further when the net of analysis is expanded to include the various non-elected boards and agencies that are increasingly a part of local governance. As these bodies are drawn into 'governing coalitions' with the overwhelmingly male upper echelons of big business, the resulting regimes can be almost exclusively masculine. Tickell and Peck (1996), for example, describe the coalition of corporate and governmental interests in Manchester, England, tied together by an elite network, self-styled as 'the Manchester mafia':

> Like the *Cosa Nostra*, this Mafia is almost exclusively male, although women are allowed into the margins of some of the families . . . There are scarcely any examples, in the Manchester scene at least, of women exercising real power in the city's new structures of governance.
>
> (Tickell and Peck 1996: 597–8)

These inequalities in local governance arise not from legal barriers but from structural conditions – including the time commitment involved, the timing and location of meetings and the expense of participation, including loss of earnings – and from cultural conditions – including the prejudices exhibited within meetings and by those responsible for recruitment – which serve to exclude and deter participants from underrepresented social groups. The determinants of local leadership can hence be reduced to three key components – control of resources, the workings of elite networks and discourses of power.

The privileges that follow from the control of some resources, such as wealth and time, are self-evident. Ownership of land can also be important in local politics, both because of the tenants and because of the development potential. Other key resources may be more subtle – for example, good communication skills are helpful in local politics and campaigning and can therefore benefit individuals with particular professional backgrounds, such as teaching (Ehrenhalt 1991; Etzioni-Halevy 1993). However, 'power' cannot be bestowed by control of any one resource, rather power is the result of different resources being blended to create a 'capacity to act'. An essential feature of urban regimes is that they bring together actors with control over key resources. Business leaders are therefore important because they can contribute financial support, and newspaper editors because they can help to shape public opinion – as can ethnic community leaders and Church ministers. As such, regimes operate through 'elite networks' that seek to connect individuals with control over or access to key resources in relatively stable relationships that can be rapidly and easily mobilised (Woods 1998a). Connections through elite networks may be important in facilitating particular policy outcomes – for example, the 1975 fiscal crisis in New York City was resolved by mediation that depended on contacts via mutual friends of the core actors (Fischer *et al*. 1977). More routinely, elite networks provide a mechanism through which the everyday interaction of local leadership can take place, serving, for example, as a vehicle for the recruitment of new actors into a regime, or for identifying appointees to political office.

The operation of elite networks can have distinctive geographical manifestations. Their membership may reflect particular spatial clusters of residence or employment, or interactions may take place within particular settings. For example, Ehrenhalt (1991) describes a picture of small-town American politics where 'a small and close-knit elite . . . made most of the important community decisions in private, frequently over coffee or lunch in a local coffee shop or restaurant' (p. xix). In Sioux Falls, South Dakota, the elite breakfasted together in Kirk's Restaurant, while in Utica, New York, the local Democratic Party boss held court at Marino's Restaurant (Ehrenhalt 1991). In these two cases, elite interaction occurred in public view; elsewhere, the 'elite spaces' may be more exclusive, serving in Goffman's (1971) terms as private 'back regions' in which the performances of the 'front region' public political stage may be contradicted (Woods 1998a).

Gaining access to the right 'elite spaces' requires matching cultural norms and expectations – having, as Bourdieu (1984) would put it, the correct 'cultural capital'. In turn, the definition of the correct 'cultural capital' may reflect local 'discourses of power' – the popularly diffused beliefs and prejudices that establish the qualities expected of leaders and define what power or influence an elite may reasonably be expected to have (Woods 1997). Thus, in Atlanta, both the 'historical families' of the old elite (Hunter 1980) and the clergy and teachers in the black community leadership were beneficiaries of status attributed through discourses of power; while in rural England discourses of power helped to maintain the elite status of farmers and landowners even as the original material basis of their power was eroded (Woods 1997). Discourses of power are often entwined with the discourses of place through which people make sense of their locality and its relation to the wider world. Such discursive associations serve to affirm the power of business leaders in industrial towns, of farmers in rural communities, and to deliver public support for boosterist policies.

Summary

Place matters in the analysis of political processes. Despite the trend of globalisation and the apparent upward concentration of power that it entails, local factors and local actors can still have political effects. The importance of place comes not from any intrinsic environmental influence, but from the distinctive configurations of social relations that exist in particular places. The discursive understandings of 'place' that people map on to these social configurations can influence both the structure of power relations within localities and the way in which residents engage politically with the wider world.

Two further issues follow from these observations. First, as places themselves are socially constructed, the ways in which they are represented and the meanings associated with them are open to contestation and conflict. As is discussed further in the next chapter, conflicts over the meaning and representation of place are frequently a focal point of local politics, as are conflicts over the ways in which place-specific interests and identities are employed to legitimise particular power structures or policies.

Second, although much of this chapter has focused on the leaders of local communities, non-elite actors are equally important in constructing and shaping the local sphere of politics. The mobilisation of an active citizenry at a local scale is examined further in Chapter 8, recognising both the role of the state in promoting community self-organising as part of a new strategy of governmentality and the interventions of grass-roots, bottom-up local protest movements.

Further reading

The subjects covered in this chapter are wide-ranging and therefore lead into a variety of specialist literature. *Theories of Urban Politics*, edited by Judge *et al.* (1995), provides an excellent overview of the different theoretical perspectives adopted in the study of local politics, including chapters on pluralism, elite theory and growth machines and regime theory. Stone's *Regime Politics* (1989) is the best contemporary study of *community politics* in North America, while the work of Jamie Peck, Adam Tickell and colleagues on Manchester provide a good British example, to which Cochrane *et al.* 'Manchester plays games: exploring the local politics of globalisation', *Urban Studies*, 33 (1996), 1319–36, and Tickell and Peck, 'The return of the Manchester man: men's words and men's deeds in the remaking of the local state', *Transactions of the Institute of British Geographers*, 21 (1996), 595–616, are accessible introductions.

The *politics of scale* are discussed in a number of places, including Marston's paper on 'The social construction of scale', *Progress in Human Geography*, 24 (2000), 219–42, Brenner's critique of this work in 'The limits to scale? Methodological reflections on scalar structuration', *Progress in Human Geography*, 25 (2001), 525–48, Purcell's commentary 'Islands of practice and the Marston/Brenner debate: toward a more synthetic critical human geography', *Progress in Human Geography*, 27 (2003), 317–32, and Cox's paper 'Spaces of dependence, spaces of engagement and the politics of scale, or, Looking for local politics', *Political Geography*, 17 (1998), 1–24. See also themed issues of *Political Geography*, 16, 2 (1997) and 17, 1 (1998), and the *Journal of Urban Affairs*, 25, 2 (2003).

Massey's article 'A global sense of place' is reprinted in her collection *Space, Place and Gender* (1994). Fiona Smith's chapter on 'The neighbourhood as site for contesting German reunification' is in Sharp *et al.* (eds), *Entanglements of Power* (2000).

The journal *Political Geography* is a good source of articles on community politics, while *Urban Studies, Urban Affairs Quarterly* and the *International Journal of Urban and Regional Research* all carry numerous papers on urban politics and power.

Contesting place

Introduction

In the previous chapter we examined how politics operate *in* and *through* places: how the particular entwinement of social and economic processes in different places can produce different policies and political strategies; and how power is distributed and exercised within communities. In this chapter we shift attention to the politics *of* place – or how places themselves can be the focus of political conflict and contestation.

A simple glance at any newspaper will provide plenty of examples of the politics of place: from globally significant conflicts such as between India and Pakistan over Kashmir, or between Israelis and Palestinians over the status of Jerusalem, to more localised disputes. Is the Alaskan tundra an environmentally sensitive wilderness in need of protection, or a commercially promising oilfield ready for exploitation? Should Spitalfields market in London be developed as a multi-million-pound office complex as an anchor for economic regeneration or is it a vital meeting place for a unique multi-cultural community?

Conflicts of this type arise because places are never neutral entities with undisputed objective meanings. Rather they are socially constructed (see Box 6.3) by individuals and groups who draw on their own experiences, beliefs and prejudices to imbue places with particular characteristics, meanings and symbolisms. Through this process many subtly different 'places' may be constructed as existing on the same territorial space. Often the coexistence is unproblematic because the different emphases are minor and do not have consequences for policy making. However, occasionally the different *discourses of place* that are mobilised are so incompatible that political conflict erupts over, for example, the appropriateness of particular developments, the legitimacy of would-be local political leaders or even the place name.

An illustration of how different perceptions about place can lead to political conflict is provided by the work of Marc Mormont, a Belgian sociologist, on one of the most culturally important 'imagined places' of the modern era – the countryside. As Mormont (1990) describes, rural areas were historically constructed as predominantly agricultural spaces. In political terms, this meant that the local power structures of rural localities were controlled by agricultural and land-owning elites, that government policy was oriented to the interests of agriculture, and that the politics of rural areas was essentially subsumed within an agricultural politics (see also Woods 1997). However, Mormont goes on to detail how processes of social and economic change in the late twentieth century disrupted this homogeneous representation of rural space, by, for example, introducing new discourses that recast certain rural places as spaces of recreation, or of conservation, or of manufacturing or service sector employment. Thus, as Mormont describes:

> there is no longer one single space, but a multiplicity of social spaces for one and the same geographical area, each of them having its own logic, its own institution, as well as its own network of actors – users, administrators, etc. – which are specific and not local.
>
> (Mormont 1990: 34)

Consequently, conflicts can arise when these different regulatory spaces adopt contradictory policies, or when the strategies pursued in the management of these spaces clash with the ideals of other imagined representations of the same territory. For instance, in-migrants to rural areas are frequently attracted by an idealised notion of rurality – the 'rural idyll' – which is sometimes tarnished by the realities of agricultural practice – noxious smells, pollution, noise and the development of functional but not necessarily aesthetically attractive buildings. In Roden, Minnesota, for example, in-migrant residents passed a local ordinance forbidding tractors to drive along the main street after 9.00 pm – frustrating the practical needs of the farming community (see also Halfacree 1994; Murdoch and Marsden 1994; Woods 1998b, 2003a). Mormont (1987) identifies such conflicts as 'rural struggles' between groups seeking to promote or defend their own symbolic representations of rurality – or, to use a more generic term, their own discourses of place.

Such experiences are not limited to the countryside. Later in this chapter we discuss a conflict over the 'gentrification' of an urban community in New York's Lower East Side, which essentially results from a similar problem of competing 'social spaces' overlapping in the same place. Other examples can be found in all kinds of geographical contexts – inner cities, suburbs, small towns, countryside, etc. – all around the world. The world in which we live has become too fluid, too interconnected and too messy for any notion of places as homogeneous, demarcated territories to be justified, and the inevitable consequence is conflict.

This chapter explores this theme from two broad perspectives. First, we look at the role of landscape in the promotion of particular discourses of place, and discourses of power within place, and at how symbolic landscapes can become the focal point of conflict. Second, we examine the interconnection between place, community and identity and how conflict over the development or representation of particular places is contested because of meanings conveyed about community identity.

Landscape and power

Landscape is the physical manifestation of place. Places as social constructs may exist as abstract ideas, on maps or in written documents, but when we actually go to a specific place, or we see a place represented in photographs, art or film, what we are experiencing is the *landscape* of that place. As such, landscapes are frequently seen as symbolic of the meanings that people attribute to particular places. By landscape we are here referring to all the various components that make up the visual appearance of a place, including the natural geomorphology, elements of cultivation such as trees, flowers, crops, gardens and parks, and the built environment of buildings, roads, paths, monuments, and so on. Thus a city centre, a factory, a theme park or a rubbish tip are all just as much a 'landscape' as a bucolic pastoral vision that we might associate with, for instance, 'landscape painting'.

Moreover, landscapes are not just assemblages of natural and manufactured objects. Cosgrove and Daniels (1988: 1) describe a landscape as 'a cultural image, a pictorial way of representing, structuring or symbolising surroundings', and if we follow this definition we can see that landscapes are full of social, cultural and political meaning.

Landscapes are *powerful* because of the role they play in structuring our everyday lives. Some of their power results from the permanence of certain landscapes, and their ability to transcend history; some results from the fact that landscapes are shared points of experience for large numbers of people who live in, work in or visit the same place. As such, points in the landscape can symbolise particular memories and meanings of place, including messages about power and politics.

We refer to landscapes that work in this way as *landscapes of power*. A landscape of power operates as a political device because it reminds people of who is in charge, or of what the dominant ideology or philosophy is, or it helps to engender a sense of place identity that can reinforce the position of a political leader. Landscapes can express power by emphasising the gap between the 'haves' and the 'have nots' – for example, through the contrasting landscapes of rich and poor neighbourhoods – and they can also become sites through which such relations of power and oppression are resisted. As Sharon Zukin (1991) remarks in her book *Landscapes of Power*, 'themes of power, coercion, and collective resistance shape landscape as a social microcosm' (p. 19).

What landscapes of power do and how they do it

Broadly speaking, we can identify four main functions of landscapes of power. First, they show who is in charge. Think, for example, of the castles of medieval Europe. As well as being important military installations, their size, construction and position served as a reminder to local people of the power of a particular baron or king. When King Edward I of England conquered Wales in the thirteenth century he ordered the construction of a series of castles, less to ensure military security – they could have little practical effect in controlling a dispersed upland population – than to symbolise the dominance of the English. More recently, 'company towns' such as Hershey in Pennsylvania, or Port Sunlight and Saltaire in England, were not just acts of social philanthropy but also served as constant reminders to the workers they housed of their complete dependence on a single company, and usually, a single industrialist (Markus 1993; Mitchell 1993).

Second, landscapes of power remind people of dominant ideologies or economic interests. An explicit example of this was the ubiquity of the red star on public buildings in communist states, but the physical layout of the landscape and the prominence of certain buildings can also convey this message. The dominance of Christian culture in Europe, for example, was historically symbolised by the centrality, size and extravagant design of cathedrals and churches; while the modern skyscrapers of the financial districts in London, New York, Chicago, Tokyo and other cities symbolise the power of contemporary capitalism (Bradford Landau and Condit 1999; Willis 1995; Zukin 1991).

Third, landscapes of power broadcast a statement about the status of a place – and send a signal to rival cities or countries. In the late nineteenth century, for example, the new industrial cities of England and Wales engaged in highly competitive programmes of public building, erecting large and elaborate town halls, commercial exchanges and libraries as symbols of their wealth, power and importance in a struggle to establish themselves as the country's 'second city'. These ambitious projects were echoed in the *grands projets* commissioned in Paris by President François Mitterrand during the 1980s and 1990s – including the Grand Arc de la Défense and the pyramid at the Louvre – aimed at reinforcing Paris's claim to be a 'global city' and France's status as a world power (Collard 1996).

Fourth, landscapes of power engender a sense of loyalty to a place, an elite or a dominant creed. We have already touched in Chapter 5 on the role played by landscape in reproducing national identity, and most capital cities have monumental spaces that perform this function. Trafalgar Square in London, for example, is an unashamed exhibition of British imperialism and military might and serves as a focal point of patriotism. At a more personal level, public statues celebrate and venerate particular political leaders and dynasties, as does, more subtly, the naming of public buildings after local or national 'worthies' (Atkinson and Cosgrove 1998; Johnson 1995; Osborne 1998).

These functions are performed by landscapes of power through architecture and through the ordering of space. Architecturally, the size, shape and building materials of particular buildings and monuments can express power in terms of command over resources, wealth and property. The architectural style used may symbolise certain discourses of power and place. For example, classical architecture is often used for government and judicial buildings because it implicitly suggests a link with the classical ideals of justice and democracy (Cornog 1988). More explicitly, Napoleon copied the triumphal arches of ancient Rome in building the Arc de Triomphe in Paris in order to identify his empire with the power and longevity of the Roman Empire. Similarly, the use of sculpture, statues, murals, inscriptions and other symbols on and in monuments and buildings can explicitly convey political messages.

On a larger scale, power is expressed through the ordering of space, for example in the central location

Plate 7.1 The alignment of the Washington monument and the US Capitol as viewed from the Lincoln monument along the Mall, Washington DC

Courtesy of Michael Woods

of royal palaces, government buildings, monuments and – at a more mundane level – factories and markets. Other monuments and buildings express power through their visibility – they are meant to be seen and to be constant reminders to the subordinate population of an elite's power. These include monuments situated on hills above towns and cities, as well as manufactured vistas such as the Mall in Washington DC (Plate 7.1), the royal Mall leading to Buckingham Palace in London and the *Grand Axe* in Paris from the Louvre through the Arc de Triomphe du Carousel and the Place de la Concorde to the Arc de Triomphe de l'Etoile and La Défense. Such landscapes are rarely constructed on undeveloped territory and power is therefore also expressed through the displacement of subordinate populations. When the Earl of Dorchester wanted to create a lake to show off his new stately home at Milton Abbas in Dorset in 1770 he moved a whole village because it was in the way (Short 1991).

Kandy: an overt landscape of power

Landscapes of power operate at different scales and with differing degrees of subtlety. One of the best documented examples of an overt landscape of power is the city of Kandy in what is now Sri Lanka (Duncan 1992, 1993). As Duncan describes, political power in Kandy at the start of the nineteenth century was based on a discourse of kingship derived from two mythical models. The first, the Asokan model, was based on accounts of a third-century BC Indian Buddhist monarch, which is a model of a righteous ruler devoted to the welfare of his people, while the second, the Sakran model, based on the model of Sakra, the king of the gods, suggests that monarchs on earth should also be universal, all-powerful god-kings. Duncan argues that both models of kingship were manifest in the landscape of Kandy – the Asokan by public works and temples, the Sakran by great palaces and monumental spaces. As the king gradually lost authority over the state bureaucracy in a struggle with his nobles he attempted to compensate by appealing to both discourses of power by building more temples and further elaborating his palace and monuments.

The result could be seen in the landscape of Kandy in 1800. The city consisted of two rectangles – the supposed shape of the city of the gods in heaven (Figure 7.1). The eastern rectangle was a monumental space, representing the city of the god-king come down to

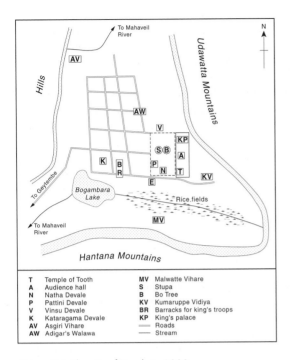

Figure 7.1 The city of Kandy in 1800

Source: Duncan (1992: fig. 4), copyright © 1992, Cambridge University Press

mounted a guerrilla campaign, cutting off the supply routes, then attacking the weakened British garrison, and eventually recaptured the city and executed the British soldiers. This victory was used by the Kandyan king, Sri Vikrama, to reassert his control over the nobility. He did this by embarking on a magnificent building programme to demonstrate his power to supporters and opponents alike.

Between 1809 and 1812 there was nearly continuous rebuilding and enlarging of the city, palace and royal gardens. Around the palace he built the Celestial Rampart decorated with an undulating pattern representing clouds – this was supposed to show the palace rising above the clouds just as the city of Sakra rose above the clouds on Mount Meru, home of the gods. The massive artificial lake to the south of the city was equally supposed to represented the Ocean of Milk at the foot of Mount Meru, while a canal cut around the perimeter of the city alluded to the annual ocean surrounding Mount Meru. Perhaps the most symbolic structure was the octagonal tower added to the temple of the tooth relic from which the king addressed his subjects. As Duncan describes:

> This octagonal structure was of great symbolic significance, for when the king stood in this tower he stood at the centre of the world with the eight points of the compass radiating out around him, symbolizing and magically reinforcing his power.
>
> (Duncan 1993: 239)

earth and paralleling in its layout the mythical palace of Sakra, including a shrine to the tooth relic of the Buddha. In the western rectangle, and around the city, were other temples and shrines. Thus as Duncan (1993) comments,

> Drawing upon the landscape models associated with the Sakran and Asokan discourses on kingship, Kandy served as a stage upon which a god-king who was also a Buddhist monarch could display both his benevolence and ritual power to his nobles and commoners.
>
> (Duncan 1993: 237)

In 1803 an English army of 3,000 men attacked Kandy. The king torched the palace, so that it could not be desecrated, and fled into the mountains with his army. The British captured the city and ransacked it, leaving a garrison to hold it, but the Kandyans

Thus as Duncan concludes:

> We can understand this building programme as an attempt by the last king to create a more perfect reproduction of the world of the gods within his capital and thereby to approximate more closely the glory and the power of Sakra, the king of the gods.
>
> (Duncan 1993: 241)

Everyday landscapes of power

Not all landscapes of power are as explicit as that of Kandy. Sharon Zukin, in coining the term 'landscape of power', applied it to the everyday landscapes of the United States – industrial towns, suburbs, shopping malls and the 'fantasy landscape' of Walt Disney World. In one memorable passage she observes that 'nowhere is the dialectic of concentration and exclusion, power and vernacular, more visible than from the elevated subway train crossing the East River between Manhattan and Brooklyn over the Manhattan bridge' (Zukin 1991: 184). On one side of the bridge, down-town New York presents a classic landscape of power as 'tall towers of steel, concrete, and glass create a layered panorama of twentieth-century finance' (Zukin 1991: 184); but for Zukin the real expression of power lies in the contrast between the skyscrapers and the more vernacular landscape glimpsed as the subway rises above ground:

> the red-brick tenements of Chinatown, built in the 1880s for Italians and Jews, testify to a still active immigrant presence. Window level with the train open on Chinese-run garment shops, while in the streets below spill stands of green cabbages and scallions, purple-skinned eggplant, and oranges.
>
> (Zukin 1991: 184)

Another illustration of everyday landscapes of power comes from Israel, where Dvora Yanow (1995) has analysed the power manifest in the construction of a community centre in a working-class neighbour-hood. The centre was designed to offer local children an 'escape' from neighbourhood life, but as Yanow describes, it also emphasises the leadership of the middle classes and represents the imposition of a middle-class worldview on the working-class community:

> The architecture, landscaping, interior design, and furnishings of the Community Center buildings tell a story of otherness and difference that has a particular ethnic and class character. It is a life-style story about Western, middle-class Israeli life that is told by people who are Western and middle-class to lower-class, development town and urban neighbourhood residents who are largely non-Western in origin.
>
> (Yanow 1995: 412)

As Yanow details, the power of the middle class (and the powerlessness of the working class) is symbolised through almost every element of the building's design, construction and location. While the architect may not have intended it to be interpreted in such a way, the community centre inevitably becomes a symbol of relative advantage and disadvantage and therefore of unequal power:

> Community Center construction materials – stone, glass, wood – were not used in residential or other public architecture in the development towns and city neighborhoods. More expensive than the local vernacular, they represent the availability of financial resources – wealth – and their associated class and social status. Interior design elements – paneling, upholstery, appurtenances – repeat this same message. The scale – the massiveness of built space – also is much larger than that of surrounding public buildings. This expansiveness is echoed in the wide approaches to the Centers, which typically also are set off on both sides and in back. . . . Scale and surrounding space tell of the command of resources. To take up space physically is also a sign of power and control: The Center stands alone, without challenge.
>
> (Yanow 1995: 411)

Because landscapes of all types can symbolise and express power in these ways, it is unsurprising that buildings and monuments in the landscape can become the focal point of conflict. This includes conflicts that surround new developments that will significantly alter the landscape of particular places, especially if the transformation concerned has a greater social or cultural significance. Ley and Olds (1988), for example, describe the political conflict that surrounded the Expo '86 fair in Vancouver, driven partly by the social impacts of developing the exposition's site, partly by the cost to public funds and partly as a debate about the cultural meaning of Expo and the place identity of Vancouver. Landscapes of power are also contested, however, through the distortion of their symbolic order. For example, graffiti artists and vandals can subvert the symbolic meaning of monuments and buildings.

The next section briefly discusses three other ways in which political power is contested through conflicts over landscape.

Contesting landscape

Monumental landscapes and the politics of memory

Discussing the 'power of landscape' with respect to the American city, Dolores Hayden has observed that:

> Identity is intimately tied to memory: both our personal memories and the collective or social memories interconnected with the histories of our families, neighbors, fellow workers and ethnic communities. Urban landscapes are storehouses for these memories, because natural features such as hills or harbors, as well as streets, buildings and patterns of settlement, frame the lives of many people and often outlast many lifetimes.
>
> (Hayden 1995: 9)

Hayden describes the encoding of the vernacular landscape with individual memories and meanings, but other features in landscape – notably monuments and memorials – are explicitly designed to structure and shape collective memories and folk memories, including historical narratives about place. However, just as history is written by the winners, so monumental landscapes are built by the winners, and the events and people they commemorate are rarely uncontroversial. One side's victories are another side's defeats, one side's heroes are another side's villains.

The interpretation of history matters because historical events and sites are used as props to support political campaigns in the present. Such is the contemporary power of monuments and statuary that the state has often been concerned to exercise control over what and who is commemorated and where.

This is illustrated in Nuala Johnson's (1995: 59) study of public statuary in nineteenth-century Dublin. Before the 1850s only two types of statue existed in Dublin – royal monuments and memorials to British

military heroes such as Wellington and Nelson. Both served to reinforce British rule in Ireland and, as Johnson notes, 'inscribed Dublin as a provincial capital within a Union whose centre was London'. As Irish nationalism grew, so did demands for the representation of Irish heroes. The first to be commemorated were literary figures, but in 1898 – the centenary of the 1798 nationalist uprising – a proposal emerged for a statue of one of the most charismatic nationalist leaders, Charles Parnell. By this time nationalists dominated domestic politics in Ireland and the Dublin Corporation supported the proposal, stating that 'no statue should be erected in Dublin in honour of any Englishman until at least the Irish people have raised a fitting monument to the memory of Charles Stuart Parnell' (quoted by Johnson 1995: 59). However, Parnell was a controversial figure, disgraced for adultery, and the proposal split Irish society. The ensuing argument – which the pro-statue lobby eventually won – played an important role as a means of debating what kind of nation an independent Ireland might be. As such, Johnson observes, 'statuary offers a way of understanding nation-building which moves beyond top-down structural analyses to more dialectical conceptualisations' (p. 57), in which the bottom-up actions of the public can be incorporated.

Conflicts over monumental space can relate not only to the subject and design of a monument, but also to its location – especially where the site itself is imbued with historic and political significance that contrasts with the symbolism of the monument. The fusion of these various sources of conflict was explored by David Harvey in his analysis of the construction of the basilica of the Sacré-Coeur in Paris (see Box 7.1).

The politics of statues and streets in Eastern Europe

In few places has the politics of landscape been as highly charged as in the former communist states of Central and Eastern Europe. The manipulation of landscape was explicitly used by the communist regimes to emphasise their power and control, and to try to generate a sense of loyalty and affinity among the public towards the Communist Party and leadership. As noted

BOX 7.1 MONUMENT AND MYTH: THE BASILICA OF THE SACRE-COEUR

The basilica of the Sacré-Coeur on Butte Montmartre is one of Paris's most prominent monuments. As David Harvey observes, 'its five white marble domes and the campanile that rises beside them can be seen from every quarter of the city' (Harvey 1979: 362) (Plate 7.2). To Parisians it has been a permanently visible symbol since its construction at the end of the nineteenth century – but a symbol of what?

Plate 7.2 The basilica of the Sacré-Coeur in Paris, a highly visible monument in the city's landscape of power

Courtesy of Michael Woods

The first clue is the name: it is a shrine to the cult of the Sacré-Coeur, the sacred heart – the idea that human guilt needs to be assuaged by offering prayers to the heart of Jesus Christ, which was pierced by a centurion's lance during his suffering on the cross. The cult had gained some popularity in eighteenth-century France. Louis XVI dedicated himself to the sacred heart and Marie-Antoinette's last prayers before she was executed were to the sacred heart. As such the cult of the Sacré-Coeur became a symbol of French monarchism. It enjoyed a second period of popularity in the mid-nineteenth century as French Catholics and monarchists were faced with the collective threats of republicanism, secularism and capitalist industrialisation. It was in this period that pressure for a shrine to the Sacré-Coeur to be built mounted, and become embroiled in the politics of the Franco-Prussian War of 1870–1 and the Paris Commune.

The Franco-Prussian War was a consequence of Bismarck's expansionist policies and his attempts to secure Prussian authority in the Rhineland. In July 1870 the Prussians invaded France and in September they besieged Paris, a siege that was to last until the following January. For many traditionalist French Catholics and monarchists the defeat was a divine punishment inflicted on a morally decadent France, and the nation needed to repent to the Sacred Heart.

However, at the same time, radicals in the besieged Paris had seized power and declared a republic, the Paris Commune. The declaration of peace in January 1871 and the election of a conservative government based at Versailles – a symbol of the old regime – only frustrated the tension, and the new French government resolved to get rid of the troublesome radicals in Paris. In March 1871 the French army marched into Paris and seized the battery of cannons on Montmartre. Crowds of working-class Parisians spontaneously set out to reclaim the cannons.

'On the hill on Montmartre, weary French soldiers stood guard over the powerful battery of cannons assembled there, facing an increasingly restive and angry crowd. General Lecomte order his troops to fire. He ordered once, twice, thrice. The soldiers had not the heart to do it, raised their rifle butts in the air and fraternised joyfully with the crowd. An infuriated mob took General Lecomte prisoner. They stumbled across General Thomas, remembered and hated for his role in the savage killings of the June days of 1848. The two generals were taken to the garden of No. 6 rue des Rosiers and, amid considerable confusion and angry argument, put up against a wall and shot' (Harvey 1979: 370).

The response of the government was to launch a full military invasion of Paris, in which 20,000–30,000 citizens were killed. For the conservatives, however, it was the two generals who were declared to be martyrs 'who died in order to defend and save Christian society' and it was as a memorial to their martyrdom that plans for the basilica of the Sacré-Coeur were finally approved.

There was some debate over its location. It was suggested that it should be built on the site of the present Opera Garnier, but eventually the hilltop of Montmarte was chosen, because it marked the spot where the generals had been executed and because it was only from there that the symbolic dominance of Paris could be assured. The site was originally earmarked for a fortress, but the Archbishop of Paris persuaded the government that 'ideological protection might be preferable to military'.

Thus by the time building started on the Sacré-Coeur it had become not only a symbol of Catholicism and monarchism, but also a symbol of atonement for the sins of modernism, a memorial to the 'martyrs' killed by the mob, a totem of national identity and an ideological fortress, reminding people of the consequences of straying from Catholic conservatism. The Sacré-Coeur took over forty years to build, and during that time its conservative associations led to many attempts to stop it, not least by the working-class residents of the districts of Montmartre and La Villette which were overshadowed by the new structure. As Harvey describes: 'the Basilica symbolized the intolerance and fanaticism of the right – it was an insult to civilization, anatagonistic to the principles of modern times, and evocation of the past, and a stigma upon France as a whole' (Harvey 1979: 379).

In 1880 a proposal was mooted to build a replica of the Statue of Liberty – being built in Paris at the time as a gift to the United States – in front of the Sacré-Coeur in order to subvert its meaning. But the proposal came to nothing, as did all attempts to stop construction. It was finally dedicated in 1919 as a symbol of French nationalism in the wake of victory in the First World War, by the then President of France, Georges Clemenceau, who, as a young man, had been one of the leaders of the Paris Commune and one of the main opponents of the basilica's construction.

As such the basilica of the Sacré-Coeur works as a landscape of power in multiple ways, but it is also a contested landscape, the construction of which was actively contested, and the meaning and interpretation of which have continued to be contested.

Key reading: Harvey (1979).

earlier, the red star symbol of communism was ubiquitously used on public buildings, and thousands of statues of Lenin and other Russian and national communist leaders were erected. Elaborate war memorials to the 'liberating' force of the Red Army were constructed, often in prominent locations such as the Gellért hill overlooking Budapest, while streets, plazas and even whole cities were renamed after communist heroes. Chemnitz in East Germany became Karl-Marx-Stadt and St Petersburg in Soviet Russia became Leningrad.

Landscape acted as an omnipresent reminder of communist power and when communism collapsed in 1990 the monuments of the old regime became the immediate casualties of political change. Journalist Tiziano Terzani, for example, describes the demolition of Lenin's statue in the Tajik capital of Dushanbe during the disintegration of the Soviet Union that followed a failed putsch in 1991:

> The execution took place at dawn: precisely at 6.35, when the first rays of the sun struck the roof of the socialist-pink palace of the local Parliament, on the square which a week ago was rebaptized 'Liberty Square'. They put a steel cable round his neck, a huge yellow crane started pulling, and Lenin, as if unwilling to leave that pedestal from which he had ruled for seventy years, slowly keeled over to one side and collapsed in pieces: the first statue, symbol of the October Revolution, to be destroyed in Soviet Central Asia. An event of great historical importance.
>
> (Terzani 1993: 251–2)

Similar scenes were repeated across the region. Outside Budapest a 'statue park' was set up as a tourist attraction to rehouse hundreds of unwanted statues removed from Hungarian villages, towns and cities. Many of the monuments that were left to stand were no longer maintained and their growing state of disrepair became an equally powerful statement of the new political order. Plate 7.3, for example, shows the Russian war memorial in the small town of Baja, southern Hungary. The memorial had occupied a prime position in the town park, but by 1996 it had become a potent symbol

Plate 7.3 The Red Army memorial in Baja, Hungary – stripped of plaques and insignia and left untended

Courtesy of Michael Woods

of the rejection of communism – overgrown and untended. By contrast, just a couple of hundred metres away new sculptures representing local Hungarian historical figures were clean and cared for.

As the communist landscape of power was dismantled, political conflicts emerged over what should replace it. One example of this was the problem of renaming streets in the former East Germany. As Azaryahu (1997: 479) notes, street names can play a similar role to monuments and memorials as 'commemorative street names conflate the political discourse of history and the political geography of the modern city. Spatially configured and historically constructed, commemorative street names produce an authorized rendition of the past.' Under communism, street names that commemorated Prussian leaders were eradicated and gradually replaced by the naming of streets after Communist Party leaders. With the end of communist rule a new round of renaming was commenced, but different cities adopted different strategies. In Leipzig, for example, historical Prussian figures were not recommemorated, but rather new names were invented that promoted the new discourse of a united Germany:

> While the theme of democratisation was articulated by decommemorating the Stalinist past of the GDR, the theme of national reunification was mainly articulated in geographical terms, most notably by naming streets after former West German cities, such as Heidelberg, Ulm, Karlsruhe, Mannheim and Heilbronn. The 'new' geography thus inscribed into the street signs meant an extension of the national territory to include both East and West Germany in the framework of a united Germany.
>
> (Azaryahu 1997: 485)

In Berlin the problem was given greater significance by the decision to relocate the capital of the reunified Germany from Bonn. In 1993 the city's senate assumed the power to rename streets in areas associated with the 'capital city function' and adopted a programme of restoring historical names that brought it into conflict with the more moderate district councils. One particular flashpoint was Otto-Grotewohl-Strasse, previously known as Wilhelmstrasse. The problem, as Azaryahu observes, 'was that this traditional name was laden with historical associations and nationalistic meanings unequivocally linked with the German Reich. A restoration of the old name, therefore, could also be understood as an attempt to imply that German reunification also meant the restoration of the Reich' (p. 487). Instead the district council proposed the name Toleranzstrasse (Tolerance Street) as a symbol of a new, non-aggressive, Germany polity. This, however, proved unacceptable to more nationalist politicians, who sought to recreate the previous global importance of

Berlin, and who launched a court challenge. Eventually the Berlin senate ruled that the street should be renamed Wilhelmstrasse.

Landscapes of control and exclusion

In the examples discussed above landscape has been used to convey symbolic power. However, the ordering of landscape can also be employed as a means of physically exerting power by restricting the movement of people, imposing divisions between groups and controlling development and standards of living. Atkinson (2000), for example, describes how one of the strategies employed by Italy to control the nomadic Bedouin population of its Libyan colony in the early twentieth century was to restrict the Bedouin's mobility by forcing them into fenced camps. The passage of arms and supplies for resistance forces across the desert from Egypt was also countered by the erection of a 282 km barbed-wire fence along the Libyan–Egyptian border. As Atkinson comments:

> Although incongruous in the midst of the Saharan landscape – particularly given the use of modern military and communications technologies – here again, Italian conceptions of fixed, impassable boundaries were eventually materialised, in this instance, by territorializing the desert interior along Italian lines.
>
> (Atkinson 2000: 115)

Other states have combined the use of physical barriers with the ordering of space through bureaucratic mechanisms such as planning laws to order and control their internal population. In apartheid-era South Africa the power of the white minority was reinforced by the Group Areas Act which spatially divided racial groups in terms of residence (Western 1996). The apartheid city was planned in a way that gave white areas every benefit in terms of aspect, weather and access to resources. In doing so, the law effectively controlled the movement of the disenfranchised black majority and denied them access to quality education and health facilities. Furthermore, putting the plan into practice meant moving existing populations, with dislocated non-white communities not moved *en bloc* but rather broken up and split between different areas, thus weakening the social ties that might form a basis for resistance. The implementation of the proposals was contested in a number of places, but only whites' objections were ever taken on board in revised plans.

Such strategies of spatial control are designed to minimise resistance and therefore permit very little internal contestation. However, challenging the assumptions and dynamics of spatial control strategy can be a tool of external opposition to contentious regimes. This is demonstrated, for example, in the work of radical Israeli architect Eyal Weizman, whose project 'The Politics of Verticality' has critiqued Israeli planning strategy in the West Bank. Through the use of 3-D maps, Weizman shows how the Oslo Accords allowed Israel to retain sovereignty over air space and the subterrain even where nominal sovereignty of the surface was granted to the Palestinian Authority – a vertical division of territory that Weizman acknowledges was a practical method to enable the two communities to put into practice claims of separate sovereignty over the same space, but which he also demonstrates has been exploited by Israel to restrict the longer-term potential of Palestinian sovereignty by, for instance, building tunnels and bridges under and over nominally Palestinian territory (Weizman 2002).

Contesting community and identity

A second way of approaching the contesting of place is to think about place not as a landscape, but as a community. As discussed in Chapter 6, 'community' is a vague and malleable term that need not necessarily have to do with place, but when we do think of 'communities of place' we are thinking about groups of people who develop solidarity and a shared identity based on an association with a particular territory. Often place association is employed to define certain characteristics of a community, so all kinds of images and stereotypes are produced and reproduced about the Scots and the Irish, about New Yorkers and

Californians, about people from different villages, or from different city neighbourhoods. As these characteristics become adopted by individuals as part of their personal identity, so individuals are moved to fiercely defend that particular representation of place. This process can be a uniting force for a community, but it can also be used to exclude certain nonconforming groups and individuals – often defined in terms of race, religion, class or sexuality – from fully participating in the community. When excluded groups choose to contest the dominant discourse, the questions of what a place means and how it is represented become a divisive issue and a source of conflict.

Discourses of place and community can be articulated in a wide range of forms. Sometimes they are encoded in the landscape – as discussed in the previous section – but they are also articulated in the writing of local history, in folk tales and folk songs, in guidebooks and postcards, by jokes, by sports mascots and the chants of sports fans, and by 'official' symbols such as coats of arms and flags. For example, one particularly contentious dispute over the representation of place relates to the use of the Confederate flag in the southern United States. Box 7.2 discusses the debate to remove the flag from the South Carolina state capital, but the flag has also provoked conflict in other states, notably Alabama and Georgia. In Georgia the issue dominated the 2002 gubernatorial elections, with the Democrat governor who had removed the Confederate cross from the state's flag in 2001 ousted from office by the first Republican to be elected Governor of Georgia since 1876.

Parades, pageantry and politics

Representations of place are also articulated through festivals, carnivals and civic pageantry. Such local rituals have performed an important function in community building since the medieval period, although many were 'invented' or 'reinvented' in the late nineteenth and early twentieth centuries at a time of great social change and instability. Significantly, the pageantry often tends to celebrate local distinctiveness and romanticised versions of the past, serving to reinforce the interests of local power elites and to

BOX 7.2 THE CONFEDERATE FLAG IN SOUTH CAROLINA

The Confederate battle flag – a blue x-shaped cross with thirteen stars set against a red background – is one of the most controversial political symbols in the contemporary United States. Dating originally from the Civil War of the 1860s, the flag was resurrected for official use in a number of southern states during the 1950s and 1960s, at the time of the civil rights movement. Webster and Leib (2001) note, 'due to its association with both the nineteenth-century Confederacy and racist post-World War II pro-segregation groups, the Confederate battle flag today inflames regional sensitivities and passions like no other symbol' (p. 275). For most African-Americans the flag is an 'icon of hate', emblematic of efforts to preserve first slavery and later segregation. For a majority of white southerners, however, the flag is 'symbolic of their ancestors' struggle, sacrifice and heroism against the perceived destructive power and tyranny of the federal government during the Civil War and Reconstruction' (p. 275).

The Confederate flag was raised above the South Carolina state capitol in the early 1960s, ostensibly to commemorate the centenary of the Civil War. Its placement apparently generated little controversy at the time, yet, as Webster and Leib note, then 'the only black presence in the chambers were "porters" who acted in both janitorial and messenger capacities' (p. 276). The flag emerged as an issue only in 1993 when the German car maker Mercedes Benz decided to locate a new plant in Alabama rather than South Carolina and an Alabamian official indicated that 'the lack of a Confederate battle flag above the state's capitol had played a positive role in the decision' (p. 277). Over the next seven years demonstrations were mounted by both pro- and anti-flag lobbies, the debate inflamed by growing racial unrest. Additionally the National Association for the Advancement of Colored People (NAACP) launched a tourist boycott of the state. As in Mississippi (Kettle 2001), economics became a key motivation behind moves to remove the flag, precisely because the flag was read outside South Carolina as a signifier of the social and political attitudes of the state.

Polls taken during the prolonged debate in South Carolina showed a clear divergence of opinion along racial lines (Webster and Leib 2001), with the two sides drawing very different interpretations of the meaning of the flag for the place identity of the southern United States. For pro-flag campaigners it was a symbol of the south as a moral, Christian society. 'It represents a time when you could walk the streets without fear. A time when the little man had a chance to make a life for his family. A time when God's law was above all else' (white Alabamian, quoted by Webster and Leib 2001: 273). For opponents the flag signified a racist society and represented the dilemma felt by many about southern identity, as expressed by one African-American writer: 'I love the South and, until quite recently, fancied myself a Southerner. Even though I was born and reared in the South and do not plan to ever leave it, I no longer believe that an African-American can be a Southerner. . . . The African-American, as imported chattel, was the South's original exile, the bastard who could not join the fraternity' (quoted by Webster and Leib 2001: 290).

The racially divided opinion among the public was replicated in the state house, thwarting early attempts to remove the flag, but the measure was eventually narrowly passed in June 2000.

Key reading: Webster and Leib (2001).

encourage distrust of change and outside influences (Woods 1999). This can become an exclusionary mechanism, either explicitly or implicitly – yet the importance of such 'traditions' to local communities means that challenges to the imagery involved are fiercely resisted. Box 7.3 describes one such contest over racial stereotypes in a carnival parade in southern Scotland.

BOX 7.3 THE PEEBLES BELTANE FESTIVAL

The annual Beltane festival in the small Scottish town of Peebles is one of many similar 'traditions' invented in the Victorian era. Over the course of a week it involves a series of events that combine ritual with carnival, including a riding of the burgh boundaries, the installation of the warden of Neidpath Castle and a fancy dress parade. As Susan Smith has observed, 'every component of the week-long festival hinges on and contributes to the social and physical boundedness of the Burgh' (Smith 1993: 293). Moreover the pageantry stressed the importance of local culture, with, for example, 'the preservation of local custom and livelihood against the threat of incomers, outsiders and unwanted social change' forming a common theme in speeches (p. 293).

One tradition was the presence in the children's parade of 'golliwogs', or stereotype blacked-up figures. In 1990, however, 'a former teacher, born in Peebles but resident in Edinburgh' (p. 296), wrote to the festival committee and to local teachers complaining that the costumes were racist and asking for them to be withdrawn. Although the committee did not accept that the characters were racist, they reluctantly agreed to the request. Publicity in the local press, however, provoked a storm of protest. Significantly objections presented the complaint as an external attack on the community of Peebles.

How dare someone who is only a visitor tell the people of Peebles and our children to get rid of our lovable Golliwogs?

I am disgusted at the suggestion that golliwogs are taken out of OUR Beltane . . . For countless years we have enjoyed OUR Beltane processions with all races represented.

A Peeblean obviously has the Beltane future closer to his/her heart than someone who has only seen a few crownings.

Our festival is timeless . . . please do not subject it to every wind of change, however specious the arguments of the opposition.

(all quoted in Smith 1993)

Letters of protest were supported by the subversive appearance of golliwog figures in the adult fancy dress parade and, provocatively, the arrival of national media interest – reinforcing locals' perception of besiegement. Thus, as Smith comments, 'the Beltane is as much about resisting marginality as about affirming the mainstream . . . the form of the 1991 Beltane symbolizes local resistance to the intrusion of English politics into Scottish affairs . . . in contesting the meanings attached to parts of the festivals by outsiders, the events of 1991 constitute a protest against the encroachment into local life of those middle class, "high" cultural values associated with the urbanized Scottish Lowlands' (Smith 1993: 300).

Key reading: Smith (1993).

Pageantry and parades can also be used by minority groups to contest dominant discourses of community. St Patrick's Day parades in North America, for example, started as demonstrations of resistance by Irish immigrants against anti-Irish local politicians, but evolved in meaning as the Irish assumed greater political power (Marston 1989). One sign of the Irish community's growing confidence was the direction of the parade through the main streets of towns such as Lowell, Massachusetts, where Marston (1989) notes, 'the St Patrick's Day parades, as they wound through the city streets, promoted a corporate awareness that it was they, the Irish, who had built the city through hard labor, and it was they who continued to maintain it' (p. 266). Peter Jackson has observed similar spatial dynamics in the Caribbean community's carnivals in London (Jackson 1988) and Toronto (Jackson 1992), yet the events remain as points of conflict over the representation of the ethnic community and its claim to place within the city (see also Cohen 1980).

Gentrification and the defence of community

In the cases discussed above the contested concept of community has been focused on symbolic representations of the community rather than on the physical environment of place itself. However, conflicts can also arise over developments in the built environment, not for environmental or planning reasons but because of the perceived harm that would be inflicted on the local community. In some instances it may be because a development would involve the destruction of an entire village or neighbourhood and the forced displacement and possibly break-up of a local community. More commonly the process is more gradual and more subtle, as property developments attract in new residents, changing the socio-economic character of the community. Probably the best documented example of this is the contestation of gentrification in New York's Lower East Side.

For decades the traditional first home to newly arrived immigrants, often living in conditions of extreme poverty and overcrowding, the Lower East Side fostered a strong sense of community and solidarity which created what many regard as the quintessential Manhattan neighbourhood. The features of community and alternative culture survive today, but the Lower East Side has also become a fashionable site for upmarket housing redevelopment, changing the social character of the population and provoking a vociferous local politics contesting the process of 'gentrification' (Box 7.4).

The area was first settled by German migrants in the mid-nineteenth century, gaining the sobriquet 'Kleindeutschland' (Little Germany). As the industrious German settlers prospered and moved uptown, particularly to the newly developing Yorkville on the Upper East Side, their place was taken by new arrivals, guided to this 'immigrant district' by the 'street birds' who met them at the Ellis Island immigration station. Prominent amongst the new settlers were Jews

---

BOX 7.4 GENTRIFICATION

Gentrification involves the redevelopment of property by and for affluent incomers to a neighbourhood, leading to the displacement of lower-income groups who are unable to afford the inflated property prices. The term was first coined by a sociologist, Ruth Glass, in 1964 to describe the renovation of working-class districts in London. Similar processes have since been observed in most Western cities as well as in many rural areas. Urban gentrification is often associated with neighbourhoods with large older properties, such as tenements, that can be easily converted into apartments but which have become run-down owing to out-migration. At the start of the process property prices are cheap compared with other parts of the city, thus allowing significant profits to be made. Classic examples include Islington in London, Society Hill in Philadelphia and Waterlooplein in Amsterdam, as well as the Lower East Side of New York. As Smith (1996) discusses, a number of different explanations have been proposed for gentrification, including cultural theories linked with changing consumption patterns and economic arguments about the benefits of urban living. Smith also examines the role of property developers and speculators in fuelling gentrification, for example by raising rents to force out lower-income residents.

Key reading: Smith (1996).

escaping persecution in Eastern and Central Europe, and the Lower East Side rapidly became the social, political and cultural centre of New York's 2 million strong Jewish community. By 1910 the section of the Lower East Side between Hester Street and Houston Street boasted over 125 synagogues in an area just a mile long and half a mile wide.

As immigration to the United States peaked in the early 1900s, the pressure on space in the Lower East Side became enormous. Tenement buildings of six or seven storeys would house up to two dozen families, in conditions that were unsanitary, vermin-infested, severely overcrowded and deprived of natural sunlight. On the worst blocks of the Lower East Side in 1910 over 1,200 people lived on a surface area of just 120m by 50m. Amid the poverty a strong sense of community and self-help developed. Trade unions were organised, relief charities were established, and adult education classes flourished. Many of the beneficiaries of this education became politically active, campaigning for better housing and public services.

The first attempt to improve conditions on the Lower East Side was made in 1890 when the worst tenements were demolished to create parks and open spaces. However, the major phase of slum clearance came during the 1930s, 1940s and 1950s, with many tenements replaced by new public housing complexes. The district continued to attract newly arrived immigrants, notably the Chinese and Puerto Rican communities – the latter settling in the northern sector they named Loisada. In 1981 over a quarter of its residents had been born outside the United States. This continuing *mêlée* of ethnic communities combined with an eastward drift of 'bohemians' from Greenwich Village following the dismantling of the Third Avenue elevated railroad in 1955 to sustain a vibrant alternative cultural and political scene in the Lower East Side, and an enduring sense of community.

However, the combination of low property prices and the newly fashionable bohemian atmosphere also appealed to property developers, who began moving in during the 1970s, buying old run-down tenement buildings and refurbishing them as luxury up-market apartments. As Neil Smith (1996) shows, the 'gentrification frontier' quickly penetrated into the heart of

the Lower East Side during the late 1970s and early 1980s. Developers even began to change the cartography of the district, renaming Loisada 'East Village' to make it sound more attractive to potential investors and buyers. The process had a detrimental effect on the community of the Lower East Side in a number of ways.

First, the redeveloped apartments were sold at premium prices beyond the reach of local residents. Second, tenants were evicted or forced out by rent increases so that landlords could redevelop property for sale or rent at much higher prices. Third, the new middle-class residents attracted up-market shops and restaurants, forcing traditional local businesses out; fourth, developers and new residents applied pressure to 'clean up' the overall appearance and atmosphere of the neighbourhood, clamping down on the homeless and others judged to be 'out of place'.

On the night of 6 August 1988 riot police moved in to enforce a curfew on Tompkins Square Park in the heart of the Lower East Side, evicting the homeless, youths, drug dealers and drug users who inhabited the park by night. The operation did not go to plan, but was resisted not only by the park users, but also by local people who saw the action as an attempt 'to tame and domesticate the park to facilitate the already rampant gentrification' (Smith 1996: 3). As Smith describes, a riot erupted, marking a key moment in the struggle against gentrification on the Lower East Side. Resistance to gentrification continues, mobilised by community groups such as the Lower East Side Collective and the Coalition for a District Alternative (CODA), who use art, direct action and participation in local politics to further their campaign. Meanwhile the interests of developers are promoted by state-sponsored public–private agencies like the Lower East Side Business Improvement District (BID) and the Southern Manhattan Development Corporation.

In 1997 a new focus of conflict in the Lower East Side emerged when Mayor Rudolph Giuliani started the process of selling off city-owned 'community gardens' to developers (see Box 7.5). The gardens' sale is particularly contentious in terms of its meaning for the local community as it not only contributes to gentrification by releasing land for development, but also

Plate 7.4 The Charas Community Center in New York's Lower East Side, sealed off and ready for development

Courtesy of Michael Woods

Plate 7.5 The last protest against the eviction of the Charas Community Center

Courtesy of Michael Woods

removes a from of semi-public space that has formed an important meeting place for the community. At the same time the city council sold the Charas Community Center, a former school that had been unofficially occupied and used by community activists since the 1970s, to a developer intending to turn it into luxury apartments (Plates 7.4 and 7.5). Through these actions the city authorities effectively weakened the

BOX 7.5 THE COMMUNITY GARDENS OF NEW YORK'S LOWER EAST SIDE

The community gardens, or *casitas*, of the Lower East Side were created during the 1960s and 1970s when residents took over derelict plots of land left by the demolition of buildings at a time when the population of the area was in decline. As well as providing pockets of green space in the midst of a densely packed urban neighbourhood, the gardens played an important social function. Collectively maintained by local people,

continued

Plate 7.6 The East Ninth Street community garden, Lower East Side, New York

Courtesy of Michael Woods

they became meeting places for residents and locations for community events. Many had noticeboards which publicised community information and political campaigns. The gardens also became an expression of community identity. While some were very simple, others were themed and some contained distinctive features such as elaborate sculpture (Plate 7.6).

Although not maintained by the city council, the gardens were technically owned by the council's Parks Department. In 1997, however, Mayor Giuliani transferred responsibility for the gardens from the Parks Department to the Housing Preservation and Development Department, with the intention of selling them off for development. The first four gardens were auctioned in July 1997, together with the Charas Community Center, and bulldozed that December. Community groups mobilised in opposition to the sell-off, concerned at the loss of valuable social spaces, and their campaign attracted considerable media attention. In May 1999 114 community gardens across New York were saved from development when they were bought by Bette Midler's New York Restoration Fund and the Trust for Public Land for a combined total of $4.2 million. However the policy of privatisation has continued and a number of other gardens remain under threat.

Key readings: For more about the case study see www.cityfarmer.org/nydestroy.html and the 'garden preservation' section of www.earthcelebrations.com.

organisational framework of the community in the Lower East Side, and with it the capacity for political mobilisation.

Summary

This chapter has demonstrated that the contestation of place is often a central element in political conflict. This arises because the meaning of place is not value-neutral. Different actors – who may be individuals or organisations – socially construct different places coexisting over the same territory, and tensions are generated when elements of the different 'imagined places' prove to be incompatible. As actors then move to promote or protect their particular 'discourse of place', political tensions can become political conflict. It can take a range of different forms and can be focused on a whole range of different expressions of place. In some cases it is the interpretation of certain features in the landscape that is at issue; in others how a place is represented through pageantry, or how it is symbolised by flags and other insignia; in yet other cases the conflict may revolve around the impact of development on the character and identity of the local community. Usually these kinds of conflict are not just about place. They are also about class or race or gender or other social divisions. But at the same time they are not entirely reducible to class or race or gender because of the significance of place in framing the dispute. It is by recognising and exploring the role of place in political conflicts of this kind that geographers can make a distinct contribution to understanding such processes.

Further reading

The material covered in this chapter leads to three different sets of literature for further reading. Harvey's article 'Monument and myth', *Annals of the Association of American Geographers* 69 (1979), 362–81, is an excellent starting point for reading about landscape and power, while Yanow's paper 'Built space as story: the policy stories that buildings tell', *Policy Studies Journal*, 23 (1995), 407–422, and Duncan's chapter 'Representing power: the politics and poetics of urban form in the Kandyan Kingdom' in Duncan and Ley (eds), *Place/ Culture/Representation* (1993), are both very accessible examples of analysis of landscapes of power.

For more on monumental landscapes see Atkinson and Cosgrove, 'Urban rhetoric and embodied identities: city, nation and empire at the Vittorio Emanuel II monument in Rome, 1870–1945', *Annals of the Association of American Geographers*, 88 (1998), 28–49; Johnson, Cast in stone: monuments, geography and nationalism', *Environment and Planning D: Society and Space*, 13 (1995), 51–66; Osborne, 'Constructing landscapes of power: the George Etienne Cartier monument, Canada', *Journal of Historical Geography*, 24 (1998), 431–58; Robbins, 'Authority and environment: institutional landscapes in Rajastan, India', *Annals of the Association of American Geographers*, 88 (1998), 410–35.

For more on power in everyday landscapes see Hayden, *The Power of Place* (1995), and Zukin, *Landscapes of Power* (1991).

The politics of parades, carnivals and pageantry are explored by Jackson in two papers, 'Street life: the politics of Carnival', *Environment and Planning D: Society and Space* 6 (1988), 213–27, and 'The politics of the street: a geography of Caribana', *Political Geography* 11 (1992), 130–51, as well as by Marston, 'Public rituals and community power: St Patrick's Day parades in Lowell, Massachusetts, 1841–1874', *Political Geography Quarterly*, 8 (1989), 255–69, and Woods, 'Performing power: local politics and the Taunton pageant of 1928', *Journal of Historical Geography*, 25 (1999), 57–74. However, the most explicit discussion of pageantry as a contested event is given by Susan Smith, 'Bounding the borders: claiming space and making place in rural Scotland', *Transactions of the Institute of British Geographers*, 18 (1993), 291–308.

The key reading on gentrification in New York's Lower East Side (and elsewhere) is Neil Smith's book *The New Urban Frontier* (1996), but see also Abu-Lughod, *From Urban Village to East Village* (1994) and Mele, *Selling the Lower East Side* (2000).

Web sites

The struggle over gentrification and the sale of community gardens in the Lower East Side of New York is well documented on the Web. A pro-development representation of the neighbourhood is presented by the Southern Manhattan Development Corporation at www.thelowereastside.org, while different perspectives can be found on sites relating to the many active campaigns in the district. The campaigns surrounding the community gardens are reported at length at www.city farmer.org/nydestroy.html, and in the 'garden preservation' section of www.earthcelebrations.com – which also includes a map of the gardens marking those that have been destroyed. A report on the Charas Community Center eviction can be found at www.tenant.net/ Tengroup/Metcounc/Nov01/charas.html.

PEOPLE, POLICY AND GEOGRAPHY

Democracy, participation and citizenship

Introduction

In this book we have approached political geography from three different starting points. In Part 1 we started with the state, its relation to territory, its use of spatial strategies and its engagement with the wider global context. Part 2 started with place, first considering the nation as a place, then the role of sub-national localities in mediating political processes and finally the emergence of conflicts over the meaning, representation and regulation of places. In this final part of the book we start with people. In particular, this chapter examines the geographies that are intrinsic to the ways in which people engage with the political process as citizens. These include the influence of place-specific factors on voting behaviour, the implications of the territorial pattern of electoral districts on the outcome of elections, the promotion of the local community as a site of 'active citizenship' and the use of place and space as a resource by protest movements.

Citizenship

Cutting across all the above themes is the concept of *citizenship* (Box 8.1). We have already touched on citizenship, either explicitly or implicitly, in a number of places in this book. In Chapter 4, for example, we discussed how work by Mark Purcell has introduced ideas of citizenship into *régulation* theory as a means of exploring the changing relationship between the state and the citizen. Notions of citizenship and citizen

---

BOX 8.1 CITIZENSHIP

Citizenship codifies the relation between the individual and the state. At one level, citizenship is a mark of belonging – our national citizenship is a sign of the nation-state to which we 'belong'. This is a legal notion of citizenship which we acquire either through birth or through application, and which then defines certain legal rights that we enjoy as citizens and certain legal responsibilities that we must perform as citizens. The right to vote and the responsibility to pay taxes are examples. At a second level, however, citizenship exists through its *practice* in ways that may extend responsibilities and restrict rights beyond the legal framework. For example, the practice of citizenship within a particular local community may be about helping that community through, for instance, various types of voluntary work. Equally, members of some minority groups may find that their *de jure* citizenship rights are in practice compromised by, for example, racist or homophobic attitudes (see Smith 1989).

Key readings: Smith (1989) and Isin and Turner (2002).

action have also been implicit in our discussion of nation building and nationalism, globalisation, community power and contesting place.

Citizenship is essentially an unwritten contract between the individual and the state which defines the responsibilities that a citizen has to the state, and the rights that they are entitled to in return. However, as T. H. Marshall (1950) noted in one of the classic works on citizenship, the rights and responsibilities of citizenship are not set to any absolute standard, but are the product of a dynamic process of social development. Marshall details, for example, how in the emergence of liberal democracy in the nineteenth and early twentieth centuries the primary emphasis was placed on the political rights of citizens, such as the right to vote or the right to freedom of speech. In post-war Europe the focus shifted to social rights, such as the right to education and to public health services, as social democratic welfare states were constructed. This was not replicated in the United States, where the economic rights of the individual remained more important – a balance that has been introduced into Europe through the processes of state restructuring of the last two decades. As such, citizenship is always politicised and contested.

The three sections of this chapter reflect some of these different expressions and experiences of citizenship. The first section, on electoral geography, focuses on one of the core rights (and, arguably one of the core responsibilities) of citizenship – the right to vote. The second section examines the promotion of an *active citizenship*, in which the state has emphasised the responsibilities of the citizen and weakened the universal nature of the social rights of citizenship. Finally, the third section looks at the exercise and defence of citizens' rights through protest, including cases in which the meaning of citizenship has been contested through challenges to the state's territorial authority and through manipulation of the symbols of citizenships – for example, by protest camps issuing their own 'passports'.

As the final section will demonstrate, citizenship is intrinsically geographical. Our citizenship identifies us with particular territorial units and the validity of the rights and responsibilities of citizenship have

spatial limits. It is therefore surprising that there has been relatively little direct engagement with the concept of citizenship by geographers. (For notable exceptions not discussed here see Parker 2002, Painter and Philo 1995 and Woods 2004.) Moreover, much of the work that has been done has not examined citizenship as an object of enquiry in itself, but rather has used citizenship as a tool in critiquing the inequalities of modern society. Following an agenda that relates to the next chapter's discussion of geography's engagement with policy, geographers have compared the *de jure* rights of citizenship that citizens should in theory enjoy to the *de facto* citizenship that they actually experience. This includes work on constraints placed on access to housing by ethnic minorities by racism (Smith 1989), and on how homophobic intimidation restricts the spaces in which gays and lesbians feel able to perform rights of expression enjoyed by other citizens (Bell and Valentine 1995; Valentine 1993).

Electoral geographies

In a democratic society the right to vote – and consequently the right to select and remove governments – is perhaps the most fundamental right of the citizen (see also Box 8.2 on democratisation). However, it should also be noted that the outcome of elections rarely reflects the pure, rational decision of the electorate. Geography keeps getting in the way. In this part of the chapter we explore the two main ways in which this happens – first when local factors influence voting decisions, and second when the geographical structure of the voting system distorts the result. We then examine the cases of two elections where geographical factors were essential in shaping the result – albeit in very different ways – the British general election of 1997 and the US presidential election of 2000.

Geographical influences on voting behaviour

The mapping of voting behaviour is one of the oldest elements of political geography, dating back to 1913

BOX 8.2 GEOGRAPHIES OF DEMOCRATISATION

The discussion in this chapter relates primarily to advanced liberal democracies in which citizens enjoy wide-ranging social and political rights, including the ability to choose (and remove) governments through fair and free elections. However, much of the world's population has no such freedoms. In over seventy states, power is exercised by unelected totalitarian regimes or a superficially 'democratic' system is restricted by the suppression of opposition parties, vote rigging, voter intimidation and controls on the freedom of speech. Since the 1980s there have been a number of high-profile instances of 'democratisation', notably in Central and Eastern Europe, South Africa and parts of Asia. These events have been positioned by some commentators as forming part of a 'third wave of democratisation' (Huntington 1991). According to Huntington's model, the 'first wave of democratisation' began in the United States in the early nineteenth century and continued to 1922, embracing the establishment of parliamentary democracies and the universal franchise in Europe, North America, Australia, New Zealand and parts of Latin America. The 'second wave' followed the end of the Second World War and lasted only until around 1962, during which time democracy was reintroduced in parts of Europe and confirmed in many newly independent postcolonial states such as India. The 'third wave of democratisation', it is argued, began with the overthrow of the Salazar dictatorship in Portugal in 1974 and continues to the present day, having reached its crest with the democratisation of Central and Eastern Europe in the early 1990s.

The democratisation of states is of interest to political geographers because, as Bell and Staeheli (2001) describe, democratisation is often conceived of not just as a historical shift, but also as a geographical process of diffusion. The task of mapping the tide of democratisation has become an industry in its own right, involving government agencies, policy think-tanks and academic researchers. Bell and Staeheli (2001) argue that in order to measure the diffusion of democracy on a global scale, mechanisms have had to be constructed to facilitate cross-national comparisons. In particular this has been done through the use of a 'democratic audit', surveying states against a checklist of political rights and civil liberties, such as fair elections and a free press. However, Bell and Staeheli contend that this approach reduces democracy to a set of procedures and institutions. For example, they note in the case of the audit used by the Freedom House think-tank that 'in attempting to evaluate the openness of a society to dissenting opinion, the survey team examines the kind of laws and institutional protections for speech, but pays limited attention to actual speech within the country' (p. 186). They warn: 'The consequences of conflating the measures of democracy with democracy itself are that it narrows the realm of possible configurations that might constitute a democratic or representative regime, and it severs the conceptual link between elections and civil society in such a way that the latter can be left to languish as a regime struggles to create a procedural democracy. The result can be the election of "illiberal regimes"' (Bell and Staeheli 2001: 188).

Thus the evaluation of democratisation is a subjective process that is inevitably informed by strategic, geopolitical, considerations. With this in mind, three further observations can be made. First, the promotion of Western-style parliamentary democracy can involve the imposition of inappropriate institutions and procedures to replace traditional forms of political organisation with strong democratic elements, such as tribal councils. Second, strategic considerations can mean that states with a poor human rights record are defended as 'democracies' while states in which free elections lead to the rise of parties that are Islamist (e.g. Turkey and Algeria in the 1990s), xenophobic right-wing (e.g. Austria in the 1990s) or leftist anti-American (e.g. Nicaragua in the 1980s) are subject to international condemnation. Third, the 'democracies' of Western states such as the United States and Britain contain flaws that can produce outcomes that might be condemned as 'undemocratic' if they were to occur elsewhere – as the discussion of the 2000 US presidential election in this chapter illustrates.

Key reading: Bell and Staeheli (2001).

when French geographer André Siegfried produced maps comparing party support in the *département* of Ardèche with the region's physical, social and economic geography. Siegfried's work was simplistic, descriptive and tended towards environmental determinism, but revealed an essential truism – that voting patterns vary spatially and that there is a relationship between them and the spatial distribution of other social and economic entities. This is not entirely surprising. In most advanced democracies the party system is based on historical social, cultural or economic 'cleavages' – for example, between classes or between religious or ethnic groups (Rokkan 1970). As these social groups tend to be geographically concentrated, the parties associated with them will also find their support varying spatially. So social democratic parties built on working-class mobilisation have historically secured more support from urban areas with a higher working-class population, while pro-employer conservative parties have attracted support from more middle-class suburban and rural districts. Similarly, the tendency of black Americans to vote Democrat is reflected in the correspondence between voting patterns and the racial composition of neighbourhoods in cities such as New York and Los Angeles.

When the spatial distribution of classes or ethnic groups shifts over time, so the associated geography of voting evolves. For example, middle-class migration from British cities to suburbs and rural areas in the 1960s and 1970s resulted in an increasing polarisation of rural–urban voting patterns – which weakened again when urban working-class solidarity was undermined by economic change in the 1980s (Johnston *et al.* 1988). However, if voting patterns simply reflected the political preferences of socio-economic groups the geography would be purely coincidental. On the contrary, electoral geographers have argued that geographical factors can amplify social biases in voting.

First, people tend to vote in a similar way to their neighbours, even if their own socio-economic status suggests that their loyalties should lie elsewhere. This 'neighbourhood effect' operates because individuals' interpretation of political news and issues is mediated through local discussion, creating a predisposition for people of all backgrounds to adhere to the dominant political narratives of their locality. For example, Butler and Stokes (1969) suggested that while 91 per cent of working-class residents in British mining districts voted for the Labour Party, only 48 per cent of working-class residents in predominantly middle-class seaside resorts voted Labour. The neighbourhood effect thesis has been criticised by some writers, who argue that such patterns can be explained by mobilisation around consumption issues such as housing and transport (Dunleavy 1979; Prescott 1972), but it has more recently been employed as an explanation of differential levels of turn-out in elections (Sui and Hugill 2002).

Second, party loyalties can be disrupted by personality politics and issue voting. A 'friends and neighbours' effect means that candidates can generally expect to poll more strongly in their home area while anomalous results can be produced when specific local issues overshadow issues in the national campaign. For example, in the 2001 British general election the victorious Labour Party lost a seat, Wyre Forest, to an independent candidate campaigning on the single issue of saving a local hospital from closure. Third, variations in the level of campaigning by candidates and parties between different electoral districts or constituencies can influence both the voter turn-out and party support (Johnston and Pattie 1997). This is becoming increasingly significant as voters behave more like consumers, selecting between different competing party 'brands' rather than following traditional class or ethnic loyalties (Sanders 2000).

Gerrymandering and malapportionment

Few governments or political leaders are elected simply on the basis of the number of votes cast. In most electoral systems the vote is effectively filtered, either by the election of representatives from geographical constituencies to a legislature or by the operation of an electoral college. In 'first past the post' electoral systems, such as those used in Britain, Canada and the United States, where the winning candidate 'takes all', this means that votes cast for a losing candidate in any constituency are effectively 'wasted'. A party that loses in every constituency by just one vote will not be represented in the legislature, while a party that wins in every constituency by just one vote will hold all the seats. As such, parties are discriminated against if their support is geographically dispersed and advantaged if their vote is concentrated in particular localities. Historically, this bias has meant that small parties with nationwide support but no 'strongholds', such as the British Liberal Party, have been underrepresented. It can also produce dramatic outcomes such as in the 1993 Canadian election when the governing Progressive Conservative Party collapsed from 169 seats to just two (see also Johnston 2002a).

Which party benefits most from biases in the electoral system will depend in part on how the boundaries of electoral districts or constituencies are drawn. For example, Figure 8.1 depicts four towns with equal populations but different divisions of support between two parties, which must be organised into two constituencies. If town A and town B are paired together in one constituency and towns C and D in the other then the Red party and the Blue party would each win one constituency. However, if town A was paired with town C, and town B with town D, both constituencies would be won by the Red party.

The deliberate manipulation of electoral district boundaries for political gain is known as 'gerrymandering' after a nineteenth-century Governor of Massachusetts, Eldridge Gerry, who authorised a peculiarly shaped electoral district in Essex County. When a local newspaper's cartoonist saw a map of the new district he accentuated its resemblance to a salamander, or as his editor named it, a gerrymander. There is some debate about the correct application of the term today (Johnston 2002a, b; Moore 2002) but no shortage of examples of fragmented, sinuous or otherwise misshapen electoral districts designed, for instance, with the distribution of racial groups in mind (Figure 8.2).

In Britain responsibility for delineating parliamentary constituencies lies with an independent Boundary Commission. However, party biases can still arise. *Malapportionment* can occur between reviews as urban-to-rural migration steadily reduces the electorate of urban constituencies while increasing that of rural constituencies. This has historically benefited the Labour Party before reviews, as the urban concentration of its support meant that it needed to poll fewer votes per constituency to win seats than the Conservatives, whose vote was stronger in the expanding rural and suburban areas; but it has also meant that the adjustments enacted by each review have immediately benefited the Conservatives (Johnston *et al.* 1999, Johnston 2002a).

Acting in combination, these various factors can have a significant effect on the result of an election, as can be seen from two contrasting examples – the 1997 British general election, where geography helped to exaggerate a landslide victory for the Labour Party, and the 2000 US presidential election, where geography was crucial in determining the winner.

The 1997 British general election

The 1997 election resulted in the end of eighteen years of government by the Conservative Party and the election of a new Labour government. The period of Conservative dominance was produced by a number of factors, but was assisted by shifting electoral geography in which Labour built up huge majorities in its core urban seats while the Conservatives' appeal to the growing middle classes enabled them to spread their vote more efficiently and win seats in rural and suburban areas and provincial towns (Johnston *et al.* 1988). In 1997, however, the tables were turned as Labour was elected with the largest parliamentary

Town A	Town B
Red party: 700 votes Blue party: 300 votes	Red party: 500 votes Blue party: 500 votes
Town C	**Town D**
Red party: 400 votes Blue party: 600 votes	Red party: 550 votes Blue party: 450 votes

Figure 8.1 Different ways of drawing constituency boundaries can produce different election outcomes

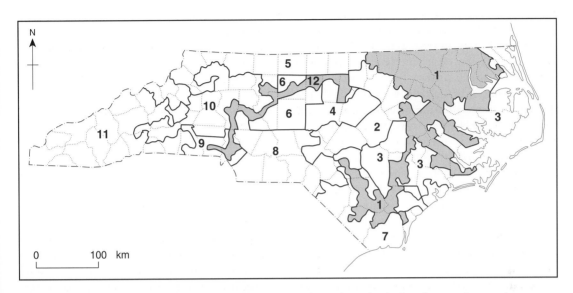

Figure 8.2 US House of Representatives electoral districts in North Carolina. The shading illustrates how misshapen districts 1 and 12 have become for partisan purposes

majority for sixty-two years, despite polling fewer actual votes and a barely higher proportion of the vote than the previous Conservative government had received in the 1992 election.

The difference in 1997 was where Labour's votes came from, and this in turn reflected a more conscious manipulation of the geography of the electoral system by both the Labour Party and the electorate. First, Labour adopted a strategy of targeting 'Middle England' – middle-class voters in the 100 or so suburban, provincial city and semi-rural constituencies that it needed to win to gain power. Both the party's policies and its campaign message were specifically tailored to appeal to this relatively small group of voters, producing a higher than average swing from the Conservatives to Labour in regions such as south-east England (McAllister 1997; Pattie *et al.* 1997). Second, Labour also concentrated its campaign activity in these target constituencies, undertaking relatively little campaigning in its 'safe seats'. The party lost some votes in its traditional strongholds to fringe parties and abstentions, but its vote was more efficiently spread, helping it to win more constituencies (Johnston 2002a). Third, voters too became more savvy about the

electoral system. In many constituencies electors voted 'tactically' for the opposition party best placed to defeat the incumbent Conservative, with Norris (1997) estimating that tactical voting and exaggerated local anti-Conservative swings were collectively responsible for the loss of forty-six Conservative seats over and above the national trend (see also Johnston 2002a; Pattie *et al.* 1997).

The 2000 US presidential election

If geography exaggerated the result of the 1997 British election, in the American presidential election of 2000 it was crucial in determining the winner. After a prolonged and contentious process the Republican candidate, George W. Bush, was declared elected with 271 of the 538 votes in the electoral college, but with nearly 540,000 *fewer* actual popular votes than his Democrat rival, Al Gore. Given the closeness of the result, geography helped to determine the outcome in three critical ways.

First, Bush polled popular votes where they counted for more in the electoral college. The President is elected not by a popular vote but via an electoral college in which each state has a designated number of electors roughly proportional to its population. The candidate who polls most votes in a state gets all that state's electoral college votes (except in Maine and Nebraska, which have slightly different systems). Although the allocation of electoral college votes is roughly proportional, there are discrepancies. California, as the largest state, has fifty-four votes, or the equivalent of one electoral college vote for every 550,000 residents. Wyoming, in contrast, has three votes, or the equivalent of one electoral college vote for every 150,000 residents. In other words, a vote cast in Wyoming has three times the value of one cast in California. In 2000 Gore won large states, including California, New York, Pennsylvania and Illinois, but Bush led in more small states where each individual vote counted for more (Figure 8.3). Thus Bush's 4.5 million votes in the thirteen small states of Alaska, Arkansas, Idaho, Kansas, Mississippi, Montana, Nebraska, Nevada, New Hampshire, North and South Dakota, West Virginia and Wyoming netted him fifty-six electoral college votes, whereas Gore's 5.7 million votes in California gave him only fifty-four electoral college votes (Archer 2002; Johnston *et al.* 2001).

Second, Bush also gained more from the efficiency of his vote spread. Both Bush and Gore won a number of states by a margin of less than 2 per cent, but whereas Gore piled up huge majorities of more than a million votes in California and New York, Bush was crucially declared the winner in Florida by 537 votes. This narrow margin brought Bush twenty-five electoral college votes – equivalent to the combined college votes of the eight smallest states – making the vote of each one of the 537 electors 1,584 times more influential than that of the average voter (Warf and Waddell 2002).

Third, Bush benefited more than Gore from geographical variations in the administration of the election. His narrow lead in Florida was assisted by a controversial state law restricting the ability of convicted felons to vote, which disproportionately discriminated against Democrat-leaning black voters, and which did not apply in other states. At a more local level, the ability of wealthier (and Republican-leaning) counties to purchase modern technology for casting

and counting votes, while poorer (Democrat) counties made do with antiquated equipment, made a small yet crucial difference in reducing errors and completing recounts in Florida before the deadline imposed by the Supreme Court.

Active citizenship and participation in communities

Voting may be the classic mode of political participation, but it is an increasingly unfashionable one. Turn-out in the 2001 British general election was a record low of 59.1 per cent of the electorate, while only 51 per cent of the voting-age population voted in the 2000 US presidential election. Commentators have offered a wide range of explanations for this disengagement, which we do not have space to cover here, but it is worth noting that one possible factor is that voters consider the question of who controls government to be less important than previously because governments are less involved in delivering services to people than before. The restructuring of the state in the late twentieth century, as documented in Chapter 4, has cut back the scope of state activity across a range of policy areas and has shifted more responsibility on to citizens themselves. This transition can be seen as the emergence of a new form of *active citizenship* in which citizenship is perceived not as something that is passively received from the state, but as something that must be actively performed by individuals through participation in governance and sharing responsibility for the defence of citizenship rights. Thus parents are expected to raise funds for schools, residents are expected to join 'neighbourhood watch' schemes to guard against crime, and communities are expected to produce their own initiatives for economic regeneration (Kearns 1995).

The promotion of active citizenship has been identified by some writers with a shift in *governmentality* – or the way in which government renders society governable (see Box 2.2). One of the strengths of the governmentality approach is the attention it pays to the *apparatuses of security*, such as health, education, social welfare and economic management systems, that

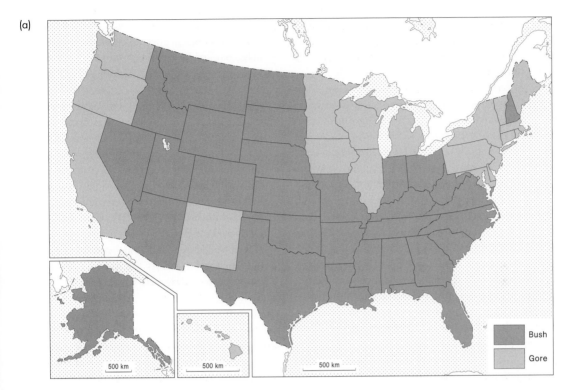

Figure 8.3 States won by George W. Bush and Al Gore in the 2000 US presidential election. (*a*) A conventional projection of the United States. (*b*) States represented proportionally to the vote in their electoral college. (*c*) States represented proportionally to the relative weight of each resident's vote

are employed by advanced liberal states in the government of a population (Dean 1999). The precise form of organisation of these systems, and the extent to which access to the services they provide is positioned as a social right of the citizen, depends upon the *political rationality* adopted by a state at a particular time. Thus it has been proposed that advanced industrial states have experienced a transition from a rationality of 'managed liberalism' – epitomised by the Keynesian welfare state – in which 'social rights' were emphasised and state planning was organised at a national scale to a new rationality of 'governing through communities' which 'does not seek to govern through "society"; but through the regulated choices of individual citizens, now construed as subjects of choices and aspirations to self-actualization and self-fulfilment' (Rose 1996: 41).

The strategy of 'governing through communities' does not necessarily mean that territorial communities have become the key units of government. The strategy can also refer to communities of interest or affiliation, such as ethnic minority communities, that need not have territorial expression; but commonly it is through geographical communities that new policy initiatives are worked, if only because such an approach suits the geographically structured apparatuses of the state. One example of this can be seen in Australian rural development policy, which in the 1990s began to emphasise the importance of building self-reliant communities that had the capacity, vision and motivation to drive their own regeneration (Herbert-Cheshire 2000). Direct state intervention in major infrastructure developments was replaced as a strategy by initiatives such as 'community builder programmes', community leadership retreats and conferences with titles like 'Positive Rural Futures'. Supporters of this strategy – and similar approaches adopted in most other developed world states – celebrate it as an *empowerment* of local communities because it removes decision making from the arbitrary processes of a distant state and places it in the hands of local citizens. However, critics have argued that it is more accurately a privatisation of responsibility:

> For those who advocate the self-help approach to rural development, its empowering potential for

rural people is a fundamental strength. In contrast, for those authors who are more critical of the underlying intentions in governmental discourses of self-help, empowerment represents little more than a rhetoric to obscure the true extent to which power remains (increasingly) in the hands of political authorities. Whichever side of the debate individual authors might take, the main issue for local people, perhaps, is not so much the intent behind discourses of self-help – that is, whether government policies are actually constructed around genuine notions of shared ownership and control or not – but rather, how those forms of empowerment are actually played out at the local level; whether individuals themselves feel empowered by the process or whether, as is suspected, it is not so much control as the added burden of responsibility that is being devolved.
>
> (Herbert-Cheshire 2000: 211–12)

As Herbert-Cheshire implies, the problem of empowerment lies at the heart of any critique of active citizenship and 'governing through communities'. The concept of active citizenship was originally informed by a notion of the *citizen as consumer* promoted by 'new right' administrations during the 1980s, which introduced a new right of choice for the citizen in their consumption of public services (see Lowndes 1995; Urry 1995). Yet consumer rights are relative to spending power, and the consumer-citizen discourse has been disempowering to those with limited resources as universal social rights have been undermined. Moreover, as the community has become positioned as the appropriate site for the practice of active citizenship, so any empowerment that might accrue to the individual has become subject to the internal dynamics of community power, as discussed in Chapter 6. In many cases, so-called 'community empowerment' has in effect been the empowerment of a community elite to dictate development strategies or organise services in ways that meet their own interests rather than those of the community as a whole (see Box 8.3; Kearns 1995). Similarly, the rationality of 'governing through communities' has produced a new geography of uneven development, as some communities have greater resources available (in terms of, for example, the wealth

BOX 8.3 ACTIVE CITIZENSHIP IN BRITISH SOCIAL HOUSING

One dimension of the promotion of active citizenship by the British Conservative governments of the late 1980s and early 1990s was the transfer of responsibility for the management of social housing from elected local councils to non-profit housing associations. This followed an earlier policy that gave the tenants of social housing the 'right to buy' their property and was aimed at reforming the management of that housing stock that remained in public ownership, as well as meeting needs for new social housing. The principle was that tenants would have direct representation on housing associations. It was argued that housing associations 'deliver accountability "downwards" to communities and are open to the influence, in practice not just in principle, of the people they serve. Moreover, they seem to be successful in delivering services which satisfy most people. . . . If the objective of developing new forms of governance is better government, then Community Ownership seems to have met with success' (Clapham *et al.* 2000: 232).

However, this rhetoric is questioned by Adrian Kearns (1992), who has examined how far the management committees of housing associations are representative of the communities that they serve. He found that only 12 per cent of all management committee members were housing association tenants, and only 49 per cent of committee members lived in any of the neighbourhoods in which their association owned property or was developing properties. Committee members were also disproportionately male owner-occupiers with professional or managerial jobs, compared with the population as a whole, while ethnic minorities and young people were underrepresented. These biases arose in part because of the expectations placed on committee members, and in part because of the methods used to recruit members of the committees. As Kearns observed, 'the popular criticism that housing associations are "self-perpetuating oligarchies" would appear, in general, to be well founded: through the use of personal recruitment methods, new committee members are likely to be from similar social backgrounds to existing committee members, are likely to hold similar views and values, and are unlikely to upset the consensus on committee' (1992: 28–9).

Kearns concludes that 'the state is shifting more resources and responsibility into the hands of voluntary and non-profit-making organisations without putting in place the necessary support arrangements to enable ordinary citizens to perform the management task in a voluntary capacity' (p. 32). In the case of housing associations this raises questions about the extent to which responsibility and power have truly been devolved to citizens, or whether a new cadre of middle-class professionals has simply been empowered. In other contexts more serious issues of territorial equality are raised. For example, Kearns (1995) questions the reach of active citizenship crime prevention schemes, compared with state provision, asking, 'Does a neighbourhood watch committee . . . feel that it is governing the entire neighbourhood or just those areas with significant participation rates?' (p. 163).

Key reading: Kearns (1992).

or professional skills of residents) than others either to provide services and facilities or to compete for funding and resources from the state.

Participation and social capital

Active citizenship and the rationality of 'governing through communities' are both premised on a strategy of increasing participation in the political process – both by drawing more people into decision-making responsibilities and by encouraging individual citizens to deepen their political participation beyond the ephemeral act of voting. Additionally, proponents of 'community empowerment' argue that the enhanced decision-making responsibility forms an incentive to civic participation. Despite these ideals, however, there

Table 8.1 Citizen participation in fifteen US cities

Level of strong participation	Type or frequency of activity	No. of respondents	%
0	Respondent participated in none of the activities	2,212	33.4
1	Respondent participated in Crimewatch, electoral campaign, contacting local officials, helping to form a new group or other activities involving some small degree of personal interaction	1,451	21.9
2	Respondent worked with others to help solve some community or neighbourhood problem	1,563	23.6
3	Respondent participated in a specific citizen group or neighbourhood association less than once a month	658	9.9
4	Respondent participated in a specific citizen group or neighbourhood association about once a month	343	5.2
5	Respondent participated in a specific citizen group or neighbourhood association more than once a month	399	6.0
	Total	6,626	100.0

Source: Berry *et al.* (1993: table 4.1)

is little evidence that such strategies have produced a greater degree of sustained participation by a larger proportion of the population than traditional modes of political participation. Research in the United States, for example, has indicated that around a third of the population are not involved in any community-based citizen action, and that the engagement of participating citizens is occasional at best (Table 8.1). Furthermore, it has been argued that citizen action has become concentrated in those areas where participants believe that there is a real, short-term, chance of achieving change – issues like 'the right to shelter; the right to worship in certain ways; to profess a sexual orientation; to carry or not carry weapons; the right not to witness pornography' (Kirby 1993: 142). As Kirby notes, this development is regarded by some as a retreat from emancipatory politics, as it allows the assertion of reactionary local ordinances while the state appears impotent: 'It may seem regressive, insofar as the state can no longer guarantee human rights, and is reduced to making promises about the length of time one

will have to wait for a hip replacement operation' (p. 142).

The American experiences in Table 8.1 are highly pertinent as the shift between rationalities of governmentality has arguably been less marked in the United States because social welfare was never developed to the same extent as in Europe, and the political and economic rights of citizens were always prioritised over social rights (Esping-Andersen 1990). As such the United States may be seen as offering a prototype for the future of active citizenship elsewhere, but it should also be noted that active citizenship US-style is not fundamentally anti-statist in that the resolution of citizen action is often achieved through local government (Berry *et al.* 1993; Kirby 1993). In so far as there has been a revival of community politics and community governance in the United States it has been the composite result of two separate processes. First, consumer-citizens have mobilised to force local governments to respond to issues connected with their own lifestyle choices. Second, the contribution of community expertise, resources and vision has been recognised as an important factor in economic regeneration.

In order to tap into latent community resources, and to stimulate citizen participation in districts where participation rates have been low, community development strategies, not just in North America but also in the European Union, Australia and New Zealand, are increasingly focused on 'community capacity building'. This means facilitating the development of networks and resources that enable communities to evaluate their own needs, identify their own solutions, access funding and implement their own projects. The role of external agencies, including the state, has become the provision of *animateurs* to supply professional advice and training. For example, a number of US and Australian states now have government-supported programmes for training 'community leaders' (see Richardson 2000). By focusing on the community scale, the 'capacity-building' strategy is intrinsically geographical, because it promotes place-based solidarity and encourages participants to emphasise the distinctness of their locality.

Some authors have identified the purpose of community capacity building as being the development of 'social capital' (see Box 8.4). The concept of social capital highlights the importance of trust and co-operation as a basis for collective action, proposing that 'the more people connect with each other, the more they will trust each other, and the better off they are individually and collectively' (Gittell and Vidal 1998: 15). It also, however, places current developments in historical perspective, as it argues that many of the social and political problems faced by contemporary society – including low levels of political participation – are the result of a weakening of social capital during the second half of the twentieth century. The concept's main proponent, Robert Putnam, has exhaustively documented evidence of the declining participation of American citizens in a vast range of collective activities – political campaigning, volunteering, membership of civic and professional associations, churchgoing, socialising with neighbours and even leisure activities such as card playing (Putnam 2000). Putnam blames these trends on changing work patterns, changing family structures and the rise of new technologies such as television which have promoted individualistic forms of leisure consumption, and argues that these have all led to a decline of trust in American society and hence of collective action.

Although Putnam occasionally disaggregates his data by state, he rarely presents any kind of geographical

---

BOX 8.4 SOCIAL CAPITAL

'Social capital' refers to the worth and potential that are invested in social networks and contacts between people. The term is intended to be analogous to 'economic capital' – or financial resources – and 'human capital' – or the skills and attributes of individuals. It has had a number of different precise usages, with some theorists, like Pierre Bourdieu, employing it at the level of the individual to describe the resources contained in an individual's social network, and others speaking of social capital in a collective sense to describe the sum value of networks and interactions in a society. Robert Putnam, the American political scientist who has popularised the term, tends towards the latter position. He further makes a distinction between two main types of social capital. The first, *bonding capital*, refers to social networks that bring closer together people who already know each other. The second, *bridging capital*, refers to contacts that bring together people or groups who did not previously know each other. Bonding social capital helps to promote community solidarity, while bridging social capital assists in enabling communities to access external resources.

Key readings: Putnam (2000) and Mohan and Mohan (2002).

analysis of social capital. However, Mohan and Mohan (2002) have identified a number of ways in which geography may engage with social capital theory. First, they propose that a geography of social capital might be explored through investigation of spatial variations in levels of participation, volunteering and the existence of organisations credited with producing social capital, linked with the development of spatially disaggregated indicators of social capital. Second, they consider the application of social capital as a concept in geographical analysis, highlighting in particular research on economic growth and uneven development, work on the effectiveness of government institutions and investigation of health inequalities. These disparate concerns are connected by Mohan and Mohan with the observation that:

> the interest in social capital results from a critique of overdetermined theorization of links between structural forces and individual experiences, a recognition that contexts matter to the outcomes of social processes, and, in particular, a critique of the excesses of free-market capitalism and failures of state intervention.
>
> (Mohan and Mohan 2002: 202)

The usefulness of social capital as an analytical tool is, however, blunted by the normative agenda that underlies Putnam's thesis. That is to say, Putnam is not content with describing how society is, but also sets out to describe how it should be – promoting policies and initiatives that encourage good neighbourliness and civic participation as a route to rebuilding social capital. For this reason, his ideas have received considerable political attention around the world. At the same, though, they have also been criticised on a number of grounds.

First, Putnam's explanations for the decline of social capital are very top-down, linked with global processes of social and technological change, and thus ignore the specific circumstances of particular communities, especially low-income and minority neighbourhoods. Similarly, his interpretation of the effect of television watching on participation has been questioned (Norris 1996). Second, his methodology has been challenged,

with critics accusing him of charting membership trends only in types of association that support his argument and pointing out that the precise way in which social capital is created through participation in often mundane activities is never made clear (Levi 1996; Mohan and Mohan 2002). Third, there is an uncomfortable disjuncture between Putnam's earlier and later work. In his earlier study of Italy (Putnam 1993) he argues that northern Italy has been more economically successful and politically stable than the south owing to the denser presence of civic and social associations in the north, but warns that the high levels of participation and trust in northern Italy have been built up over centuries and cannot be easily replicated. However, in his later work Putnam seems to want to critique the pattern of civic participation in the United States over a much shorter time scale, and to propose a short-term programme for restoring social capital. Finally, Putnam's analysis of the decline in political participation in post-war America has been challenged by critics who point to the growth of protest movements and of radical forms of citizen action – as we discuss in the next section.

Protest and citizen action

Not all forms of political participation are in decline. As conventional methods, such as voting or performing civic duties as local office holders, have lost some of their popularity, so non-conventional methods, such as protest, direct action and radical citizens' action, have grown in prominence. In the autumn of 2002 over half a million people from around the globe gathered in Florence, Italy, for the first World Social Forum – an event that combined workshops and seminars with protest marches and demonstrations in promoting an agenda of global peace and justice that was intended as a counterpoint to the annual meeting of business leaders and politicians at the World Economic Forum. The Florence gathering built on earlier mass demonstrations at meetings of the World Trade Organisation in Seattle and Genoa as part of an emergent antiglobalisation movement. Significantly, all three events involved large numbers of young people – precisely the population group that has become most detached from conventional politics.

On a broader scale, the latter part of the twentieth century also saw a massive expansion of pressure-group politics. While Putnam (2000) recorded the downward trend of participation in civic and voluntary organisations in the United States, many campaigning pressure groups have over the same period enjoyed a rapid growth in membership, especially groups associated with issues such as the environment and human rights. Friends of the Earth, for example, has grown from a handful of founder members in four countries in 1971 to over a million members in sixty-eight countries by 2002. The participation of many of these members will be fairly passive, and some will be 'cheque book activists' whose only involvement will be to make regular donations. However, the environmental movement has also been at the forefront of pioneering direct action protests that no longer rely on the state to take action on behalf of the citizen but which engage citizens in direct confrontation (see McKay 1998; Wall 1999).

The kinds of political participation described here have prompted a reworking of citizenship in three ways. First, demonstrations and direct action are an assertion of citizens' right to protest, but they also represent a lack of trust in the ability of conventional politics to respond to citizens' demands. Second, actions such as the demonstrations in Florence, Seattle and Genoa are a response to the seeping upwards of power away from nation states to transnational institutions and corporations – from the World Trade Organisation to McDonald's. As conventional politics, exercised through the apparatuses of the nation-state, offers no route of direct democratic engagement with these new centres of power, new forms of citizen action have developed to circumvent the state in challenging transnational institutions and corporations directly as acts of 'global citizenship'. One notable example of this was the dismantling of a McDonald's restaurant in southern France by activists in the Confédération paysanne led by José Bové. When Bové was charged with criminal damage he effectively inverted the court proceedings into a trial of globalisation by calling environmental and social justice campaigners from around the world as witnesses (Woods 2003b).

Third, radical action of this kind has proved that active citizenship does not just occur on agendas set by the state – citizens themselves can define the scope and purpose of their engagement by embarking on radical action to protect and promote communities. As such, non-conventional politics are not just about protest, but can combine advocacy with mutuality, especially around issues that are peripheral to mainstream conventional politics, such as gay rights and AIDS activism.

Integral to these processes has been a reconfiguration of the spaces through which citizenship is performed. On the one hand this has involved an up-scaling to draw together participants from across the world to target transnational institutions and corporations about issues of global significance in acts of global citizenship. Yet, at the same time, such protests are always situated and the ways in which they interact with, manipulate and subvert the geographies of their location can impact on their success (see Box 8.5). Linked with this

BOX 8.5 AIDS ACTIVISM IN VANCOUVER

One example of the reworking of citizenship through radical action is Michael Brown's study of AIDS activism in Vancouver. As Brown (1997a) describes, citizens' responses to the AIDS epidemic took a number of forms, from self-help to political protest, all of which operated in a local political climate dominated by socially conservative city and provincial governments that were reluctant to engage with the growing crisis faced by AIDS sufferers. At one level, AIDS activism involved a form of active citizenship as volunteers provided care

continued

and support for sufferers, often providing services that were not available through the state. At another level, however, *caring voluntarism* could not be separated from the political objective of raising awareness of the AIDS problem. This included the exhibition of a memorial quilt, consisting of 12 ft by 12 ft patchworks commemorating people who had died from AIDS-related causes. Contrary to some theories of radical democratic citizenship which have emphasised the importance of confrontation, Brown (1997a) argues that the quilt was equally a political act of radical citizenship because it served to convert grief and mourning into a political statement in a highly visible public space. The transgression of public space was also significant for the more confrontational protest activities of a radical group, ACT UP (the AIDS Coalition to Unleash Power). Formed in 1990, Vancouver ACT UP aimed to raise awareness and demand positive action from government and health institutions through public stunts and demonstrations, including a die-in and the occupation of the Health Minister's office. Significantly, although the state was nominally the target of ACT UP's protests, the majority of its actions took place not in state-centred spaces but in the public spaces of civil society – including theatres and shopping streets. As Brown (1997b: 157) comments, 'the state was rarely engaged on its own turf; rather it was challenged in public spaces of civil society. Civil society, operationalized in public space, was a weapon of resistance against the state.' This strategy helped to achieve some objectives of raising public awareness, but it also, Brown argues, contributed to the ultimate failure of ACT UP. The group's life span was short, organising just nine events over two years, and it proved less successful than other more mainstream AIDS organisations in influencing the city government. Brown (1997b) identifies the reasons as connected with geography in two ways. First, he argued that ACT UP misunderstood the nature of local politics in Vancouver. It was essentially a US model transplanted to Canada which failed to realise either that the more generous welfare system in Canada limited the scope of its demands of the state, or that local political culture articulated through the city's newspapers would be antagonistic to its style of protest. Second, in adopting a political geography of protest that attempted to set civil society against the state, ACT UP failed to appreciate the extent to which the local state and local civil society were interwoven. As Brown (1997b) notes, 'key state bureaucrats were secretly funding AIDS service organisations, much to [the ruling party's] chagrin. In other words, by attacking the state *from* civil society, ACT UP failed to acknowledge the ironic linkages between the two spheres, which were common knowledge in local AIDS politics' (p. 163). Thus, while radical democratic citizenship may be oriented around global issues such as AIDS, its actions remain situated and continue to need to be sensitive to locality.

Key readings: Brown (1997a and b).

---

is a new localism, in which local communities have been identified as the starting point for radical action – an approach encapsulated in the slogan of the environmental movement of 'Think global, act local'.

New social movements

The rise of non-conventional politics has promoted the emergence of not just new forms of political action and citizenship but also new forms of political organisation. Whereas conventional politics has been organised through the formalised structures of political parties, trade unions, trade associations and lobby groups, the groups that make up the environmental movement, or are involved in AIDS activism or anti-globalisation protests, tend to be less formal, decentred and often ephemeral in their existence, with little or no direct connection with the electoral process. As such they have been characterised as 'new social movements' – a designation that is designed both to distinguish them from other forms of associational activity that do not have political objectives, such as social and voluntary organisations and religious movements (see Box 8.6), and to emphasise that they are the product of a paradigm shift in politics.

Social movements themselves are not new – the labour movement being a prime historical example. However, the distinction of 'new' social movements is used to indicate a shift in emphasis in the objectives of social movement activity:

> While the 'old' social movements emanated from the class structure of industrial capitalism and aimed at uprooting the material inequalities produced by the mode of production, the 'new' social movements cut across classes and are guided by non-material considerations.
>
> (Fainstein and Hirst 1995: 183)

According to Claus Offe (1987: 73), this transition reflects a wider shift between political paradigms – from an 'old' paradigm centred around issues of economic growth and distribution and characterised by action through formal representative organisations, party political competition and corporatist bargaining to a 'new' paradigm centred on issues such as the preservation of the environment, human rights, peace and social justice, in which action is informal and spontaneous, operating through protest politics, with demands formulated in predominantly negative terms. For Offe (1985) social movements have been key agents in this transition, critiquing the institutional assumptions of representative democracy. Instead, he argues, they have forged a new radical politics, expressed through a critical ideology of modernism and progress, defence of interpersonal solidarity against bureaucracies, and new forms of political organisation (see also della Porta and Diani 1999: 12). As della Porta and Diani summarise, in Offe's view new social movements are characterised by 'an open, fluid, organisation, an inclusive and non-ideological participation, and greater attention to social than to economic transformation' (1999: 12).

Thus the motivation for new social movement mobilisation cannot be reduced simply to material gain, but may concern the achievement of symbolic goals or the defence of symbolic resources. Despite the contention in some analyses of social movements that individuals will participate in collective action only when the benefits that accrue to them as individuals outweigh the costs (Olson 1968; see also Fainstein and Hirst 1995), the implication of Offe's new paradigm is that 'new' social movements have emerged in which the individual material gain to the participant is not

---

BOX 8.6 SOCIAL MOVEMENT

Diani (1992: 13) defines a social movement as 'a network of informal interactions between a plurality of individuals, groups and/or organisations, engaged in political or cultural conflict on the basis of a shared collective identity'. There are four key components to this definition. First, a social movement comprises a number of different, independent, groups who share a common purpose, but may also on occasion adopt contradictory policies or strategies and be drawn into conflict with each other. Second, the links between the component groups are informal – there is no centralised leadership or command. Third, social movements are engaged in political activity. This distinguishes them from social clubs, voluntary groups and religious organisations. Fourth, the uniting force for social movements is a shared identity, not just shared interests. Together, these components allow social movements to be distinguished 'from various forms of collective action which are more structured and which take on the form of parties, interest groups or religious sects, as well as single protest events or ad hoc political coalitions' (della Porta and Diani 1999: 16).

Key readings: della Porta and Diani (1999) and Diani (1992).

obvious – but where the costs of participation are balanced by the aim of achieving some greater good. One might think here of environmental protesters, of demonstrators against armament sales to undemocratic regimes and of campaigners against Third World debt.

As these last three examples indicate, the shift in the object of political engagement between the old and new paradigms of politics also has a spatial expression. First, because the formal electoral process has become less significant in the practice of politics by new social movements, and because many key concerns of new social movements are about global issues, the nation-state has become less important as the scale of action, with social movements often campaigning across national boundaries. Second, there has been a concurrent shift in the spaces of political action. Whereas the spaces of political action by 'old' social movements are the sites of state and economic power – parliaments, government buildings, factories, workplaces – the spaces employed by new social movements are the spaces of social and environmental power and spaces of consumption and communication – supermarkets and shopping malls, carnivals, fairs and public spaces. In part this shift reflects the fact that social movements are not interested in assuming power, in taking over government: rather they seek to change political practice and policy. A key weapon in this respect is public opinion, and crucially, winning over public opinion means enrolling the media:

> for the most part social movements use forms of action which can be described as disruptive, seeking to influence elites through a demonstration of both force of numbers and activists' determination to succeed. At the same time, however, protest is concerned with building support. It must be innovative or newsworthy enough to echo in the mass media and, consequently, in the wider public which social movements (as 'active minorities') are seeking to convince of the justice of their cause.
>
> (della Porta and Diani 1999: 183)

In order to meet the need for media coverage, many protests by new social movements are symbolic in nature rather than directly threatening to power holders, thus practising a form of political action that Paul Routledge has labelled 'the postmodern politics of resistance'.

Postmodern politics of resistance

Routledge regards the key difference between older and newer forms of political action to be the way in which contemporary struggles are 'postmodern' in their extensive mediation and symbolic nature. Postmodern politics aims not to capture the state apparatus, but rather to resist and restrict state power (and corporate power) when the exercise of that power is perceived to threaten valued environments, communities or ways of life: 'Postmodern politics, then, are characterized by heterogeneous affinities that coalesce in particular times and places as activist assemblages. Eschewing the capture of state power, they nevertheless pose challenges to the state' (Routledge 1997a: 373). The challenges to the state tend not to be mounted directly, but to be mediated through society and the media, relying on symbolic action to distort and subvert social and spatial orders:

> such a politics mounts symbolic challenges that are extensively media-led in order to render power visible and negotiable, and to attract public attention. Such a politics, and the spaces within which, and from which, it is articulated are frequently hybrid in character and ambiguous in practice and effect.
>
> (Routledge 1997a: 372)

Routledge (1997a) demonstrates this model through the example of a protest against the construction of a new motorway, the M77, through Pollock Park on the edge of Glasgow, Scotland. The protest campaign was organised by Glasgow Earth First! and involved both local people and participants who came specifically to join the protesters from elsewhere. It involved the occupation of the construction site with a semi-permanent protest camp as well as other tactics such as protest marches and sabotaging machinery. However, as well as forming a physical obstacle to the road building, the protest mounted a series of symbolic challenges to the planning of space that had led to road proposals, to the cultural norms of a car-dependent society and to the power of the state. These symbolic challenges involved both the subversion of spatial order and the subversion of citizenship – the latter being articulated by the declaration of the protest camp as a 'free state' and the issuing of its own 'passports':

> The Free State represented the 'homeplace' and the focus of the resistance against the M77, articulating an alternative space that occupied symbolic and literal locations. It acted as a place where people who were interested in the M77 campaign could learn more and get involved. . . . The Free State stood as a critique of the environmental damage caused by road building and an example of how people might live their lives differently. Its politics of articulation interwove ecological, cultural and political dimensions.
>
> (Routledge 1997a: 366)

These messages were reinforced by the mixture of symbols, icons and images created and employed within the camp (see Plate 8.1):

> In addition to the totems and tree houses – themselves hybrid sites of habitation and tactical forms of protection for trees – the Free State comprised a mixture of symbols. Abandoned cars were used to create dramatic sculptures such as 'Carhenge'. A flag of the Lion Rampant girded the trunk of a tree near to the entrance of the Free State, next to which was an Australian aboriginal land rights flag. A wind-powered generator supplied power to a portable television and stood above a mobile phone. Next to images of Celtic knots flew Buddhist-style prayer flags strung from the trees, on which the phrase 'Save our dear green place' was block-printed.
>
> (Routledge 1997a: 367)

In these ways the Pollock protest camp communicated a message about a global issue – the environment – through the manipulation of a specific site. To do so it drew together in a unique, place-specific, combination

Plate 8.1 The 'carhenge' sculpture at the Pollock Free State protest camp
Courtesy of Paul Routledge

cultural symbols and signifiers from around the world that associated the site with a plethora of struggles from renewable energy to Aboriginal land rights. In a different study, of resistance to a military base in India, Routledge (1992) describes a similar combination of physical and symbolic challenges through the manipulation and subversion of space – but drawing strength from the rearticulation of local representations of place (Box 8.7). Both examples, however, take us back to an observation that we borrowed from Michel Foucault at the beginning of this book. Not only is space, as Foucault noted, 'fundamental in any exercise of power' (Rabinow 1984: 252), but space is also fundamental in any resistance to power.

BOX 8.7 SPACE, PLACE AND RESISTANCE: THE BALIAPAL PROTEST

In 1984 the Indian government announced plans to build a missile testing range in Baliapal, a rural district in north-eastern Orissa on the Bay of Bengal coast. Baliapal had been selected as the site for the range because of its geographical isolation, climate and topography, but the proposed development posed a threat to the local agricultural economy and would have displaced some of the region's 100,000 residents. A movement was formed to oppose the development by two local politicians, one from the right-wing Janata party, the other from the Communist Party. The resistance movement transcended class and caste divisions, uniting participants around a sense of community. It used the traditional local social structure in forming a committee of traders and village council leaders and adopted a strategy of non-violent resistance that 'spanned methods of intervention, non-co-operation and protest, and persuasion' (Routledge 1992, reprinted in Agnew 1997: 231).

At one level the tactics of resistance involved physically obstructing the construction work. Barricades were erected to stop government officials from entering the area: 'In order to warn people of approaching government vehicles, conch shells were blown and *thalis* (metal plates) were beaten, thereby summoning thousands of villagers to the barricades. Once there the villagers lay down in the road, forming human road blocks' (p. 232). By restricting access to government officials, the movement subverted spatial order by challenging the territorial authority of the state. This was also done by 'demolition squads' who acted to demolish the new model villages being built for the evicted peasants.

At a second level, the Baliapal protest also mounted a symbolic challenge through the discursive mobilisation of a sense of place. As Routledge records, 'the Baliapal movement was informed and motivated by a potent sense of place which refined and strengthened the economic motivation provided by the locale. This sense of place was epitomized by the movement's ideology of "Bheeta Maati" (our soil) articulated as "our soil; our earth; our land". As one activist remarked to me: "For Baliapalis the land is our mother; our earth; our home. This is in the hearts of the people"' (Routledge, in Agnew 1997: 233).

Through these tactics, Routledge argues, 'the emergence of a people's movement in Baliapal and the adoption of a potent variety of non-violent tactics has transformed the padi fields and jungles of the area into a terrain of resistance. . . . The terrain has been articulated socio-politically within civil society where the government's legitimacy to make policy decisions which are antithetical to local community interests has been challenged by a withdrawal of peasant consent. The terrain has been articulated culturally in the form of songs, poems, the *Vichar* process, various religious idioms and the movement's non-violent tactics – those "little tactics of the habitat"' (Routledge, in Agnew 1997: 235).

Key reading: Routledge (1992).

Conclusion

This chapter has explored three means of citizen engagement with the state in a democratic society. First, through the ballot box as part of the electoral process; second, as active citizens enrolled in civic participation at a local scale; third, through involvement in protest events and social movements. Each of these activities is itself a result of particular *rights* enjoyed by citizens in a liberal democracy – the right to vote, the right to participate in government and the right to protest – but which are routinely suppressed in states where democracy is restricted or non-existent. The struggle for democracy can be cast in geopolitical terms, providing a framework for the 'new world order' and counterpoising 'Western' democracy against 'non-Western' autocracy. Such a model, however, invites a critique, not least of the ways in which the meanings of 'democracy' are constructed and of the limitations of democracy within supposedly 'democratic' states. Even in advanced liberal democracies, *de jure* citizenship rights may not be comprehensively implemented in practice, such that the *de facto* citizenship of certain groups becomes restricted – particularly for ethnic and sexual minorities (Bell 1995; Kofman 1995). The difference between *de jure* and *de facto* citizenship frequently has a spatial manifestation – on finding that their rights are restricted in particular (often public) spaces, excluded groups create and colonise more private, marginal, spaces in which greater freedom can be achieved and the enforcement agencies of the state or of an intolerant majority may be evaded (Bell 1995; Pincetl 1994; Valentine 1993).

Indeed, citizenship has become an increasingly important concept for political geographers in part because the performance of citizen rights and responsibilities is so strongly geographical. As noted in this chapter, spatial structures can shape the patterns and outcomes of political participation by citizens, as in the effect of electoral district territories on election results or in the differing opportunities for civic participation that exist in different localities. At the same time, citizens' political participation can create new geographies and new spaces. Moreover, space can be a focus of struggles over citizenship, particularly around issues of public and private space. Citizens' rights may be indicated by the extent to which citizens have full, unfettered access to and use of public space, and by the extent to which the state is limited in its right to enter or regulate private space. In liberal democracies the testing of these spatial rights revolves around concerns such as the surveillance of public space (Fyfe and Bannister 1996; Goodwin *et al.* 2000; Koskela 2002), formal or informal regulation of the use of public space by young people (Malone 2002) or the homeless (May 2000), racial discrimination in access to housing (Smith 1989) and the policing of sexual

practices in private space (Bell 1995). By contrast, the absence of citizens' rights in totalitarian states is manifested in restrictions on the use of 'public' space and in the routine invasion of private space by state agents. Resistance activities may therefore involve the challenging of these spatial controls, by, for example, holding illegal demonstrations in public spaces – as in Wencelas Square in Prague during the Czechoslovak 'Velvet Revolution' of 1989 and, more tragically, in Tiananmen Square, Beijing, in the same year – as well as through the creation and manipulation of covert 'spaces of resistance' as sites of organisation and mobilisation (Routledge 1997b; Watts 1997).

Furthermore, citizenship is geographical in that it connects political actors with particular territorial entities. The discussion in this chapter has primarily focused on citizenship as defined in terms of the nation-state, albeit sometimes articulated within local-scale communities. There is, however, no intrinsic reason why citizenship should be identified with the nation-state and citizenship, as a concept, has been associated with different territorial scales at different points during history (Isin 2002). In the contemporary era, 'national citizenship' is under challenge both from below, with the reassertion of locally constituted 'citizenships', and from above, as the sovereignty of the 'nation-state' is eroded by globalisation (see Chapter 4). The latter includes challenges posed by issues of immigration and asylum, and by multiculturalism within states (Kofman 2002; Yuval-Davis 1999, 2000), as well as by the promotion of supranational forms of citizenship, such as European citizenship within the European Union (Painter 2002). Within this context, authors such as Yuval-Davis (1999, 2000) and Painter (2002) have posited notions of 'multi-layered' or 'multi-level' citizenship in which citizenship is defined and articulated by engagement with different scales of political authority and with a range of other social identities. Thus as Yuval-Davis (2000) observes:

> very often people's rights and obligations to a specific state are mediated and largely dependent on their membership of a specific ethnic, racial, religious or regional collectivity, although they are

rarely completely contained by it. At the same time, the development of ideologies and institutions of 'human rights' means that, ideologically at least, the state does not always have full control of the construction of citizenship's rights, although it is usually left for states to carry them out.

(Yuval-Davis 2000: 171)

Yuval-Davis's reference to human rights points to a further recent development, the promotion of a notion of 'global citizenship'. This concept implies that all people across the globe should enjoy common 'citizen's rights', but also that the world population as a whole has a responsibility to the global environment and society. Thus global citizenship on the one hand involves the establishment of institutional mechanisms to define, monitor and enforce human rights, and on the other hand is a motivating force for transnational social movements such as the global environmental movement, the fair trade movement and the anti-globalisation movement (Routledge 2003). Transnational movements, in turn, both engage with local and regional citizens' groups (Perreault 2003) and require locally situated sites of articulation in which their political activities are performed (Routledge 2003). In these ways the spaces and scales of political mobilisation are constantly being remade and reconfigured.

Further reading

For reading on citizenship and political geography the best starting points are Susan Smith's paper 'Society, space and citizenship: human geography for the "new times"?' *Transactions of the Institute of British Geographers*, 14 (1989), 144–56, and the 1995 themed issue of the journal *Political Geography* on 'spaces of citizenship'.

There is a dearth of good up-to-date literature on electoral geography, especially work written from an American perspective, although Johnston, 'Manipulating maps and winning elections: measuring the impact of mal-apportionment and gerrymandering', *Political Geography*, 21 (2002a), 1–31, covers many of the key ideas. Johnston

et al. *A Nation Dividing?* (1988), provides a comprehensive but dated historical account of British electoral geography. For analysis of the 1997 British general election the best starting point is the special issue of the journal *Parliamentary Affairs*, 50, 4. For the 2000 US presidential election see papers by Clark Archer and others in *Political Geography*, 21, 1 (2002).

For reading on active citizenship see the two papers by Kearns, 'Active citizenship and urban governance', *Transactions of the Institute of British Geographers*, 17 (1992), 20–34, and 'Active citizenship and local governance: political and geographical dimensions', *Political Geography*, 14 (1995), 155–75. There are numerous papers on various aspects of community development, with the journals *Urban Studies, Urban Affairs, the Journal of Rural Studies* and *Rural Sociology* being good places to look for examples.

Two edited books provide good starting points for further reading on the geographies of protest and resistance: Sharp *et al*. (eds), *Entanglements of Power* (2000), and Pile and Keith (eds), *Geographies of Resistance* (1997). See also Routledge's paper 'The imagineering of resistance: Pollock Free State and the practice of post-modern politics', *Transactions of the Institute of British Geographers*, 22 (1997a), 359–76, and George McKay's edited book *DIY Culture* (1998).

Web sites

There are numerous resources on the Internet detailing election results and analysis which can be drawn on for research on electoral geography. For American elections Leip's on-line Atlas of US Presidential Elections, http://uselectionatlas.org/, is an excellent source of information not just on the 2000 election but also for elections back to 1824, with detailed results, analysis and interactive maps. The US Elections Central site, http://www.multied.com/elections/index.html, is aimed at high-school students but provides very accessible resources on state-by-state results as well as information on the functioning of the US electoral system. The British Politics Pages, http://www.club.demon.co.uk/Politics/elect.html, have detailed results, maps and analysis of national and local elections in Britain since 1983. For the really serious, extensive but accessible British election statistics and information is also archived on the United Kingdom Election Results site, http://www.election.demon.co.uk/.

Finally, the Political Studies Association provides links to elections-related Web sites from around the world at http://www.psa.ac.uk/www/election.htm.

Public policy and political geography

Aberystwyth, August 2003

Political geography has always had degrees of relevance to, and influence over, real-world issues (House 1973). Sometimes this has not been progressive or productive. The military dictator General Augusto Pinochet, for instance, was trained as a political geographer and used this background to remake Chile during the early 1970s. According to David Harvey, 'Pinochet did not approve of "subversive" academic disciplines such as sociology, politics and even philosophy'; geography was his poison for instilling patriotism, regulating culture and undertaking social engineering (Harvey 1974: 18). As president of the military Junta, Pinochet overthrew a democratic and elected government and undertook brutal reforms on, among other things, health and social policy. Harvey talks about the ways in which military control allowed Pinochet to smash the actors and institutions of the progressive Allende regime, which created the space for re-establishing the 'old geography' of a centralised and dictatorial power base.

At the time of writing, the relationship between geographers and public policy (defined in Box 9.1) is being critically questioned and has 'air time' in leading academic journals (see Dorling and Shaw 2002; Martin 2001; Massey 2000; *Scottish Geographical Journal* 1999). In 1999 Jamie Peck wrote an editorial statement in the journal *Transactions of the Institute of British Geographers*. This was partially a response to Brian Berry's (1994) call for more public policy analysis in geography, and also an attempt to provoke a similar debate to that raised by Harvey and others in the 1970s. Peck asked why geographers were not involved in the policy-making process under the Labour govern-

ment, given Britain's 'unique insights into "real world" processes and practices' (Peck 1999: 131). Replies followed, and the early twenty-first century is a fruitful time to consider if, or how, political geographers can contribute to public policy. If geographers are, as Doreen Massey claims, working 'themselves up into quite a lather' (Massey 2002: 645) then it could be important.

This chapter considers the importance of these debates for political geographers by discussing the different exchanges between human geography and public policy over especially the last thirty years. Although the majority of these debates have taken place in British geography journals and have often not involved leading political geographers, they have implications for the wider discipline of political geography. The chapter starts by looking at links between geography, empire and public policy, and questions whether this was the Golden Age for geographers and the policy process. The chapter then considers the 'relevance debate' of the early 1970s, which pushed public policy back on to the geographical agenda. It also discusses the more recent exchanges that are recommending a new 'policy turn' in the discipline. To think about what political geography could offer this debate, the chapter concludes by suggesting that the capitalist state's role in the policy process is a key missing link throughout these exchanges and we discuss how political geography students could consider this in their own work. By the end of the chapter it becomes evident that public policy questions social science itself (Blowers 1974). Our position on where political geographers can perhaps make a difference is summarised as Box 9.2.

BOX 9.1 PUBLIC POLICY AND GEOGRAPHY

According to Ron Martin, public policy involves 'any form of deliberate intervention, regulation, governance, or prescriptive or alleviative action, by state or nonstate bodies, intended to shape social, economic or environmental conditions' (Martin 2001: 206). Ron Johnston offers a similar synopsis, where pubic policy is the 'study of and involvement in the creation, implementation, monitoring and evaluation' of public initiatives (Johnston 2000: 656).

Key readings: Johnston (2000) and Martin (2001).

BOX 9.2 POLITICAL GEOGRAPHY AND PUBLIC POLICY

The contemporary political geographer is *not* a policy maker: they can only offer academic analysis on public policy, but in doing so can influence those involved in policy-making processes. They 'cannot take decisions' (Blowers 1974: 32). That said, we perhaps have a moral obligation to take public policy seriously in our ongoing academic analysis. This involves not only writing critically about public policy and the policy-making process. Where possible, geographers can use their various insights to inform political and policy practice, accepting the tensions between political/policy action and academic/intellectual critique. What do *you* think?

Key readings: Blomley (1994), Blowers (1974), Harvey (1974) and Massey (2000).

Geography and empire: the Golden Age of public policy?

The example of Pinochet can be contrasted with the more balanced interventions made by Sir Halford Mackinder. As we suggested in Chapter 1, the early intellectual foundations of political geography rested on the transition from systematic to regional geography, which at that time made descriptive connections between physical spaces (natural regions) and social and political ('ethnographical') worlds (Mackinder 1902). Political geography at the time was very much an inductive science and it influenced British thinking on the ascendancy of the territorial state, set within a context of rapidly shifting power relations. Mackinder was central to this context and after being Director of the London School of Economics and Political Science

– which has historically positioned academics close to Britain's national political machinery – time was spent as High Commissioner for South Russia. Chapter 1 highlighted that Mackinder's book *Democratic Ideals and Reality*, an interpretation of world power politics, was presented as a warning to the peacemakers at Versailles (Mackinder 1919).

Was this the Golden Age of close relations between human geographers and policy processes? Mackinder was heavily involved in the Royal Geographical Society, which at that time had close links with the political system, and held numerous positions in commerce and industry. Mackinder certainly 'had the ear' of Ministers. Others followed in this line and acted as advisers to parliamentary committees. Sir Dudley Stamp worked closely with the government on mapping agrarian trends and influenced post-war land-use policy. Stamp was later rewarded for his contributions to the 'use of the land' (Dickinson 1976: 7). John House discusses the involvement of geographers in urban and regional planning since the 1930s, both through the involvement of academics in giving policy advice and also through the direct employment of geography graduates within the state's apparatus. Geographers were the driving forces at the Barlow Commission's 1937–40 inquiry into the 'Distribution of the Industrial Population' – a key moment in the evolution of the Keynesian welfare state (see Chapter 4) – and occupied strategic positions within the civil service (House 1973). Brian Robson saw these interventions as effective ways of getting the human geography voice heard (Robson 1972).

After 1945 human geography searched for a stronger intellectual and scientific identity, and this pushed political and social necessity into the background. In Britain, for instance, key ties between academia and policy making were also restructured through the formation of the Institute of British Geographers (breaking away from the Royal Geographical Society). Combined, this created the space for a 'new style' of geography (House 1973). Instead of serving the needs of empire and its Colonial Surveys, and the nation and its Regional Surveys, academics developed a university-based intellectual and pedagogic discipline that went hand in hand with nurturing human capital in accordance with national socio-economic needs and modernist political priorities (see Harvey 1974; Unwin 1992). Human geography also broadened its intellectual reach by incorporating developments in, and having a critical dialogue with, other disciplines (such as sociology, economics and politics) to capture multiple ways of interpreting 'geographical worlds'. The 'old style' geography of human–physical–state interactions (House 1973), which policy makers could perhaps understand, was gradually replaced by a diverse and intellectually stimulating set of agendas.

The rise and fall of relevance debates

During the early 1970s a new 'wind of change' swept across academia and brought with it a more 'radical geography' (Berry 1972) that was not based on traditional concerns with location, classification, regularity and conformity. For David Smith and others airing their thoughts at the Association of American Geographers conferences, radical geography meant a politicised 'social geography' that practised 'social responsibility' with greater professional involvement in welfare rights, social justice, and political activism' (Smith 1971). The context of this was the Vietnam War, student riots in Paris and growing urban poverty and social inequality (see Watts 2001). Michael Dear also suggests that some geographers at the time became alienated by a geography overly focused on quantitative techniques and under-concerned with real-world issues (Dear 1999). To cut a long story short, advances in social philosophy appeared to be bringing with them a renewed sense of academic and political responsibility.

The watchword of these times became 'relevance' – introduced to gauge the degree to which geographers were making a contribution to the analysis and resolution of economic, environmental and social problems (Prince 1971). Such claims were contested, and the 1970s witnessed a wealth of debates, with clear disagreements between liberals, humanists, Marxists and others on *how* to tackle geography and public policy. In one intervention Brian Berry argued that 'an effective policy-relevant geography involves neither the blubbering of the bleeding hearts nor the machinations of the Marxists. It involves working with – and on – the *sources of power and becoming part of society's decision making apparatus*' (Berry 1972: 78, our emphasis). Berry added that academic analysis had to be interdisciplinary and follow a problem-oriented approach, where 'the solution to social problems would be facilitated by careful, clear projection of policy objectives, programme alternatives and underlying economic and social forces, proceeding towards a solution through experimentation and feedback guided by theory and analysis' (ibid.: 80). This challenged the purely intellectual and scholarly pursuit of human geography, by advocating a more applied discipline.

This created ripples across the Atlantic at the 1974 annual meeting of the Institute of British Geographers. In a presidential address Terry Coppock argued that, amongst other things, strategic research – utilising computer technology to allow prediction through modelling – was required to demonstrate the geographers' role in formulating alternative public policies (Coppock 1974). For Bridget Leach, this debate provided an opportunity to discuss how policy problems become politically constituted. Leach argued that 'diversionary tactics' were used by policy elites to protect the legitimacy of the political system and by highlighting such tactics academics could inform debate by empowering opposition groups (Leach 1974). Open questions, however, remained as to how this could be achieved, given that – with the exception of David Harvey's interventions – little attention was paid to uncovering the 'decision-making apparatus' and 'sources of power' under capitalism.

For Harvey, understanding the (public policy) world was about using this insight to change things and 'before geographers commit themselves to public policy, they need to pose two questions: what kind of geography and what kind of public policy' (Harvey 1974: 18). The two categories are not separate – they are seen as linked through the moral obligations that geographers have to create a better society. According to Harvey 'relevance in geography was not really about relevance (whoever heard of irrelevant human activity?), but about whom research was relevant to and how it was that research done in the name of science (which was supposed to be ideology-free) was having effects that appeared somewhat biased in favour of the *status quo* of the ruling class of the corporate state' (Harvey 1974: 23). Harvey's Marxist stance encouraged geographers to break out of this loop and challenge the 'corporate state' within capitalism.

Last, and perhaps most interestingly for Peter Hall, geography had much to learn from political science for uncovering the actors, power networks and organisational dynamics at work in the policy process. Hall suggested that a *new political geography* was on the horizon and:

From this, certain central lines of research seem to follow. The new style of urban political geographer, for that is what he [*sic*] seems destined to become, will be concerned with the values, the organization, and access to power of groups. He will analyse the relationship of these groups to the decision-making machinery (and the personalities who operate this machinery) at different levels of government. He will study how different agents in the decision process – politicians, bureaucrats, technicians, opinion-formers – interact, how they form alliances and coalitions, how they bargain, promise or threaten each other to obtain objectives. His concern . . . is to analyse what happens, not to postulate what should happen. Yet, by the very fact of exposing the way decisions are taken in practice, I would expect and hope that the political geographer would provide powerful suggestions for future improvement.

(Hall 1974: 51)

Hall's ideas were taken forward by geographers in their work on 'urban managerialism', which was interested in the roles played by different agents (such as local government, central government, builders, estate agents and landlords) in producing cities. Some of this research featured in a special issue of *Transactions of the Institute of British Geographers*, where Robson argued it captured 'a clear reflection of the social concern and the interest in process rather than form which have made geography in the middle 1970s a very different animal from that of a decade ago' (Robson 1976: 1). Simon Duncan's paper on 'social geography' and the city, which questioned how housing systems worked and who benefitted from them – so that an 'information base' could be provided to influence 'political will' and 'achieve change in the allocation of houses and housing resources' – was indicative of these process-based concerns (Duncan 1974: 10).

The challenge to managerialism contained some of the explanations for the reduced interest in public policy. Hall's 'new political geography' was continually challenged by those who felt that it could not sufficiently explain the links between the structure of the housing market, the actions of agents and institutions, the spatial arrangement of the city and wider capitalist society (see Bassett and Short 1980). But, instead of theorising the links between the capitalist state, class, power and urbanisation – which could be fruitful ways of capturing the policy dynamics within the contemporary city – some Marxists saw this agenda as ultimately reproducing the 'corporate state' by working within the constraints of capitalism, as opposed to transcending capitalism itself (Martin 2001).

The cooling of 'relevance' debates was further aided by the reaction to these trends. So the argument goes (Martin 2001), the 1980s gave birth to humanistic geography and then the recasting of the 'social' within social and cultural geography – both countering structurally determined processes under capitalism (as expressed by certain Marxists) and models of expected behaviour (as predicted by spatial/regional scientists). As these authors grappled with new ways of exploring the action, movement and experience of individuals within geographical settings (see *Area* 1980), attachments to public policy became increasingly tenuous. To be fair, public policy was not the object of analysis for these scholars, who were more interested in 'place' and the geographies of 'everyday life'. Some continued to work within the tradition of old-school 'social geography'. Robson, for instance, developed an applied urban geography that influenced those formulating British urban policy (see Robson *et al.* 1994). Berry followed a similar career path in the United States, through research on housing (see Berry 1994).

Trends in academia were not the only explanation for this movement away from what could be considered 'relevant' geography. The onset of neoliberalism (see Chapter 4) in especially North America and Britain alienated scholars from undertaking policy-relevant research. Policy makers required research that was ideologically relevant for justifying privatisation, deregulation and public sector restructuring, and they turned to economists practising predictive and normative thinking. Economists were the perfect ideological bedfellows for the New Right governments of the 1980s and 1990s.

Public policy in the twentieth century: shallow, deep or just grey?

This relationship between economists and policy makers is one of two entry points for more recent debates (Peck 1999). The other relates to the increasing dominance of postmodernist thinking and, according to critics, the perceived irrelevance of these approaches to real-world issues (Martin 2001). We consider each of these in turn.

The first agenda is clearly evident in the debate initiated by Peck, concerned by the fact that few geographers appeared to be advising Britain's New Labour Party (Peck 1999). Peck's work at this time centred on welfare-to-work – a key political strategy in the first term of Blair's government which, through geographically specific modes of policy transfer, sought to move Britain towards a North American (post-welfare) model (see Chapter 4). The key advisers in the United States and the United Kingdom were right-wing labour-market economists who manipulated the local embeddedness of policy, and Peck bemoans the roles that these 'intellectuals' play in globalising welfare state restructuring. Given the unique insights that geographers possess, Peck asks: where are the geographers in this policy process; why do economists have 'the ear of the Minister'; and why are geographers always at the bottom end of the policy research ladder, examining hard outcomes and being excluded from policy formulation (Peck 1999; also Massey 2001)?

Peck's explanation points to contemporary academic practices, which privilege abstract scientific knowledge over and above more practical and policy-oriented concerns. Because public policy research is not considered scientific or of 'top-drawer' quality it is deemed to be 'bad science'. This process is augmented by targeted research that can gain universities academic excellence and their staff promotion. Consequently, geographers are being encouraged to come up with theoretical innovations and write 'big papers' and through time this has reduced public policy research to a 'grey', boring and somewhat second-rate academic practice (Peck 1999).

BOX 9.3 BEYOND GREY GEOGRAPHY: SHALLOW AND DEEP POLICY ANALYSIS

Shallow analysis

This is policy research that is confined to addressing the 'stated aims and objectives' of policies from within an orthodox theoretical position. This often serves the needs of the policy-making system, which it often takes for granted, by licensing quick-fix solutions. Shallow researchers are often closer to the policy-making process than their deep colleagues. Examples of shallow researchers include mainstream economists.

Deep analysis

This is research that sees policy as politicised and contested and questions the 'parameters and exclusions of policy making'. Deep policy researchers often take a theoretically unorthodox position and question the local embedded and path-dependent nature of public policy. Examples of deep research include those undertaking critical investigations of the policy-making process.

Question

Which group is conducting the most effective form of policy analysis?

Key readings: Peck (1999, 2000).

Peck challenges these assumptions and argues that public policy research does not have to be this way: it can be theoretically and politically progressive. Policy research, then, 'is a legitimate, non-trivial, and potentially creative aspect of the work of academic geographers, but one that we are currently neglecting and/or undervaluing' (Peck 1999: 131). To take forward this agenda and its potential creativity, Peck makes a critical distinction between 'shallow' and 'deep' public policy analysis (see Box 9.3) and argues that geographers have much to offer in a deep approach that engages 'critically and actively with the policy process itself' (Peck 2000: 255).

In a reply, Jane Pollard and colleagues argue that the involvement of geographers in public policy is far greater than that implied by Peck (Pollard *et al.* 2000). They agree that geographers are generally not that involved in national-level social policy debates, but they are involved in other policy areas which have merit and cannot be written off as trivial. This argument is extended by Mark Banks and Sara MacKian, who urge Peck to take stock of the ways in which geographers are involved in evaluating the themes of 'renaissance', 'partnership' and 'social capital' within British urban policy (Banks and MacKian 2000). Peck's response emphasises the importance of teasing out relationships *between* different levels of public policy, to overcome what is seen as the danger of falling foul of the rhetoric of localism (Peck 2000). The challenge for geographers, then, 'is to connect together the smaller pictures with the bigger pictures of the policy process, to connect the specific with the general, without undermining the integrity of our particular take on the policy process' (Peck 2000: 257).

The approach adopted by Ron Martin is somewhat different and situates the decline in public policy research within a broader academic context. Like Peck and others, Martin also feels that human geography is exerting little influence on policy. Rather than questioning the motives behind what we publish, or exploring the general crisis in the social sciences in relation to 'relevance' (see Massey 2001), Martin pursues an internal critique of the discipline. Martin argues that 'much of what is done under the banner of human geography is unlikely to be seen by policy-makers as being remotely germane to policy issues' because it 'has little practical relevance for policy; in fact, in some cases, one might even say little social relevance at all' (Martin 2001: 191). For Martin, this is why the opinions of geographers are not being sought when it comes to consultations on policy making. We talk the wrong language, are lost 'in a thicket of linguistic cleverness', ask the wrong questions and generally do not deliver research findings that have 'relevance to real-world issues' (Martin 2001: 196; see also Martin 2002).

Martin takes issue with the contributions being made by postmodernism and the 'cultural turn' to human geography. As we suggested in Chapter 1, this has been influential in political geography and has sensitised us to the need to consider textual and discursive strategies. Critical geopolitics, for instance, has demonstrated the usefulness of this methodological approach (see Chapter 3). In Martin's opinion, however, concern with identity and culture has diverted our attention from the larger social and political problems of today. This is partly explained by the cultural turn's denial of 'extra-discursive reality' – it disengages itself from material power relations, and thereby does not accept that such forces also shape identity and politics (Martin 2001: 196). Consequently, in Martin's opinion, the cultural turn does not engage with the many processes and practices that provide the scenery for social interaction. It does little to challenge the 'structured determinants of sociospatial problems and inequalities' (Martin 2001: 201). Is this a fair criticism?

Martin also challenges the research designs that leading geographers use to demonstrate their claims. Too often our theories are being put into practice with a 'lack of rigour': geographers rely too heavily on selective quotations from a limited number of individuals, located in particular geographical locations (Martin 2001: 197). This in turn leads to 'fuzzy conceptualisation' (Markusen 1999) – our claims do not stand up to serious scrutiny or interrogation. Linked with this, the policy and political implications that flow from human geography are redundant and this reinforces a lack of political commitment. Do you agree with their concern?

Levelling criticism is easy and the 'difficult part is suggesting what needs to be done, how we should move forward' (Martin 2001: 202). We have pulled together some of Martin's suggestions for doing a 'new geography of public policy' in Box 9.4. As you can see, there are a number of key challenges facing geographers

BOX 9.4 TOWARDS A 'NEW GEOGRAPHY OF PUBLIC POLICY'?

'There is no single, all-encompassing, universally superior or commonly agreed theoretical framework or methodological approach on which to base our research. Thus there can be no single approach to policy analysis, no blueprint for how geographers should integrate public policy into their research or how they should evaluate its sociospatial impacts. There are different forms of, and approaches to, policy analysis ranging, for example, from the critical analysis of policy discourses and practices to reveal their underlying ideological, and instrumental content, to extensive empirical analyses of policies to evaluate their intended impacts and unintended consequences, to intensive ethnographic type investigations of precisely how particular policies affect specific individuals, groups and localities. Each provides a different "cut" on policy, and different policy issues will require different methods or combinations of methods. Public policy analysis has to be pluralistic, not monistic. We need more interesting and imaginative ways of combining qualitative and quantitative analysis, and of integrating intuition into our research methodologies and analyses. Above all, for a policy turn to occur in the discipline, our research has to become much more "action-based". We need to see research not simply as a mechanism for studying and explaining change, but – by following our investigations through to their implications for possible policy intervention and action – as instigator of change, as an activist endeavour . . . The geography of public policy is not just about evaluating policy impacts. Important though that role is, geographers should also be engaged in fundamental debates over the direction of society, economy and environment, and what policies would be required to achieve different outcomes. But, equally, it is surely as important to research and campaign for achievable reforms as it is to debate ideal transformations which have little prospect of being implemented' (Martin 2001: 202–3).

Key reading: Martin (2001).

tackling this debate and these can be summarised as: developing intellectual cohesion through practical social research; finding imaginative ways of combining qualitative and quantitative data to ensure rigour; and using 'action based' approaches to influence the direction of policies and their outcomes. Students of political geography will have to make up their own mind as to whether they agree, or disagree, with Martin's critique. What do you think? Is this debate important? What can political geographers contribute to this agenda?

Towards 'deeper' engagement with the policy process

We conclude this chapter by addressing what could be considered a missing link in this debate – namely the policy process itself, which is linked with how we conceptualise the state's changing institutional forms, functions and modes of intervention. According to Christopher Ham and Michael Hill, analysing policy making depends on some appreciation of the institutionalisation of power and representation in society, which in turn requires some understanding of the state under capitalism (Ham and Hill 1993). Rather than finding ready-made answers from within our discipline, political geography might benefit from adopting a 'post-disciplinary' stance on the state and its politics, whereby a dialogue is opened up with, and ideas are drawn from, social and political science.

To engage 'critically and actively with the policy process itself' (Peck 1999) political geography could offer an insight into the 'deeper' political arena. We could focus on how government responds to and represents its wider social environment. Public policy is not just political; it also has profound impacts on society by framing socio-spatial relations. Indeed, the two go hand in hand – the social and the political are mutually reinforcing, constructed and embedded in each other. By understanding the social situations and politics that go hand in hand with forms of state intervention and the multiple terrains through which this occurs, political geographers could begin to understand what makes public policy tick, why changes take place, and we can also begin to highlight access points for those individuals and campaign groups wishing to practise 'activism'.

The public policy debate, then, warrants a reconsideration of the capitalist state. Murray Low (2003) has suggested that the state remains an important missing link in contemporary political geography. As we suggested in Chapter 2, the state is everywhere and nowhere: it is the backdrop to almost everything that we experience. In the public policy debate, however, the state rarely makes more than a cameo appearance.

One way forward, and acknowledging Martin's (2001) point that there are many 'cuts' into the cake, is to take a *régulation* approach to public policy. As we suggested in Chapter 4, the *régulation* approach presents the state as a complex and broad set of institutions and networks that span both political society and civil society in their 'inclusive' sense (Gramsci 1971; Jessop 1997c). From this perspective, state intervention, state functions and public policy concerns relate to the 'micro-physics' of power.

Bob Jessop's 'regulationist state theory' is interesting in thinking about the links between political geography and public policy. Jessop draws on the work of state theorists and political activists such as Antonio Gramsci, Nicos Poulantzas and Claus Offe to think about the changing institutional forms and functions of the capitalist state. For Jessop, the state needs to be thought of as 'medium and outcome' of policy processes that constitute its many interventions. The state is both a social relation and a producer of strategy and, as such, it has no power of its own. State power in relation to the policy process relates to the forces that 'act in and through' its apparatus. According to this view, attempts to analyse the policy process need to uncover the strategic contexts, calculations and practices of actors involved in strategically selective, or privileged, sites (Jessop 1990a). This can be summarised as a framework that demonstrates 'systems analyses' for the undertaking of 'systematic' forms of public policy analysis (Ham and Hill 1993) – drawing attention to the intricate links between actors and forms of representation, institutions and their interventions and practices, and the range of policy outcomes available. This connects with our argument in Chapter 1 that political geography recognises intrinsically linked entities – power, politics and policy, space, place and territory.

Box 9.5 details the six dimensions of the state that appear in much of Jessop's work on the institutional forms and functions of political economy. Three are

BOX 9.5 RESTATING THE POLICY PROCESS

Institutional relations within the political and policy system

- *Representational regime.* This has a concern with delimiting patterns of representation and the state in its inclusive sense. It uncovers the territorial agents, political parties, state officials, community groups, para-state institutions, regimes and coalitions that are incorporated into the state's everyday policy-making practices.
- *Internal structures of the state.* This is the institutional embodiment of the above and it underscores the distribution of powers through different geographical divisions and departments of the state and its policy systems. This not only allows research to study the apparatus of central government; it also explores the ways in which political strategy helps to create sub-national spaces and scales of policy intervention and delivery. For political geographers, the relationship between the different politically and socially charged scales of governance is important.
- *Patterns of intervention.* This is associated with the different political and ideological rule systems that govern state intervention, such as frameworks of rights and responsibilities, the balance between the public and private, and the perceived roles of the social partners in the policy process. Additional concerns can include the discourses of citizenship, social inclusion/exclusion, universal versus targeted and selective service provision, and equality versus allocation through competition.

Wider social relations and civil society

- *Social basis of the state.* This consolidates the representational regime through civil society, which can be spatially selective, and explores the different ways in which uneven development is mobilised into the political system through targeted state strategies.
- *State strategies and state projects.* This brings some overall coherence to the activities of the state, its forms of intervention, and its policy-making priorities. The state is seen here as a political strategy, and its various policy and power networks can privilege some coalition possibilities over others and some interest groups over others.
- *Hegemonic project.* This mobilises the state and its multifarious policy-making networks and coalitions, and also tries to externalise/resolve conflicts that can disrupt policy systems, around an ideological programme of action. It thereby considers the ways in which collective action, forms of knowledge and discourses become codified and mobilised to advance particular interests. There are links here with notions of governmentality (see Chapter 2).

Key readings: Jessop (1990a), MacLeod (2001) and Peck and Jones (1995).

associated with institutional relations within the political and policy system. Jessop adds a further three to tease out the ways in which the state interacts with its wider social environment. The capitalist state, then, can be viewed as a strategic and relational concern, forged through the *ongoing* engagements between state personnel, institutions and public policy implementation. This perspective could assist political geographers and their students to delve deeper into the policy process itself. What kind of political geography for what kind of public policy?

Further reading

For further reading on the public policy debate in geography see Martin, 'Geography and public policy: the case of the missing agenda', *Progress in Human Geography*, 25 (2001), 189–210; Peck, 'Grey geography?' *Transactions of the Institute of British Geographers*, 24 (1999), 131–5; Berry, 'Let's have more policy analysis', *Urban Geography*, 15 (1994), 315–17; Coppock 'Geography and public policy: challenges, opportunities and implications', *Transactions of the Institute of British Geographers*, 63 (1974), 1–16; Harvey, 'What kind of geography for what kind of public policy?' *Transactions of the Institute of British Geographers*, 63 (1974), 18–24; Dear, 'The relevance of postmodernism', *Scottish Geographical Journal*, 115 (1999), 143–50; Dorling and Shaw, 'Geographies of the agenda: public policy, the discipline and its (re)"turns"', *Progress in Human Geography*, 26 (2002), 629–46; House, 'Geographers, decision takers and policy matters' in Chisholm and Rodgers (eds), *Studies in Human Geography* (1973).

For arguments on activism and political relevance see Blomley, 'Activism and the academy', *Environment and Planning D: Society and Space*, 12 (1994), 383–5; Tickell, 'Reflections on "activism in the academy"', *Environment and Planning D: Society and Space*, 13 (1995), 235–7; Castree, 'Out there? In here? Domesticating critical geography', *Area*, 21 (1999b), 81–6; Blowers, 'Relevance, research and the political process', *Area*, 6 (1974), 32–6; Massey, 'Practising political relevance', *Transactions of the Institute of British Geographers*, 24 (2000), 131–4; Blunt and Wills, *Dissident Geographies*.

If you are interested in finding out more about the policy process see Ham and Hill, *The Policy Process in the Modern Capitalist State* (1993); Burch and Wood, *Public Policy in*

Britain (1989); Offe, *The Contradictions of the Welfare State* (1984); Lindblom, *The Policy-making Process* (1968).

For further reading on the strategic-relational approach to the state see Jessop, *State Theory* (1990a), and 'Institutional re(turns) and the strategic-relational approach', *Environment and Planning A*, 33 (2001), 1213–35; MacLeod and Goodwin, 'Space, scale and state strategy: rethinking urban and regional governance', *Progress in Human Geography*, 23 (1999b), 503–27; Jones, 'Spatial selectivity of the state? The regulationist enigma and local struggles over economic governance', *Environment and Planning A*, 29 (1997), 831–64; Peck and Jones, 'Training and Enterprise Councils: Schumpeterian workfare state, or what?' *Environment and Planning A*, 27 (1995), 1361–96; Goodwin, 'The changing local state', in Cloke (ed.), *Policy and Change in Thatcher's Britain* (1992).

Acknowledgement

This chapter is a shortened version of a paper by Martin Jones, 'Human geography and public policy: reflections on recent and not so recent debates', mimeograph, University of Wales Aberystwyth.

Postscript

As we write this postscript the city of New York is marking the second anniversary of the destruction of the World Trade Center towers in the Al-Qaeda terrorist attack of 11 September 2001 (briefly discussed in Chapter 3). The tragic, murderous, events of that day and their consequences have cast their shadow over political geography as over so many areas of life. Toal and Shelley (2004), in a review of the state of political geography at the start of the new century, describe the response of the United States to the attacks as providing new definition and clarity to a new geopolitical order that was already emerging in the wake of the Cold War. They highlight the contradictions of the post-9/11 era, including the vulnerability of advanced techno-scientific systems, the obsession with absolute invulnerability that sits uneasily with the dynamics of capitalist globalisation, and the lack of self-reflection about domestic threats to security. Political geography, Toal and Shelley suggest, has a role in posing and pursuing questions about an unsure and insecure future.

The legacy of 9/11 goes beyond geopolitics. Smith (2002: 98) reveals the attack on the World Trade Center as an event of multi-scalar significance: 'a global event and yet utterly local' that was constructed as a national tragedy. The nationalisation of the tragedy permitted the articulation by the US government of its new geopolitical vision – the launch of the 'war on terrorism' and the targeting of the 'axis of evil' that led American, British, Australian and allied military forces first into Afghanistan and then into Iraq. The continuing problem of pacifying and controlling Iraq, with occupying troops confronted by regular attacks, has led the US administration to seek greater international assistance in the 'reconstruction' of the country, only to meet with suspicion rooted in the diplomatic friction with European governments that preceded the war. Farther east, US military personnel are still present in Uzbekistan and other former parts of the old Soviet Union, having been stationed there as part of the attack on the Taleban regime in Afghanistan. The 'new world order' of the twenty-first century is already looking very different from that of the last century. Yet the new discourse of security has its local and domestic-scale manifestations too – the closure and increased control of public space around 'sensitive' buildings, heightened levels of surveillance, and the physical and psychological abuse directed at ethnic communities, especially Arab communities, compromising their *de facto* citizenship rights (Bayoumi 2002).

In New York itself the local-scale consequences have become less explicit over time but remain present. On 12 September 2001 *The Times* newspaper of London remarked that 'the political and social geography of [New York] seems destined to change' (p. 1). These changes have also been both physical and psychological – in the use of space in the city, in the meaning of landscape and in the political opinion and behaviour of its residents (Sorkin and Zukin 2002). As we write, controversy continues about the redevelopment of the World Trade Center site, resonating with earlier conflicts over the politics of memorialisation, but also indicating a return to 'business as usual' in the politics of urban development.

Elsewhere on 11 September 2003 there is another crisis in the Middle East peace process, as the Israeli government threatens to expel the Palestinian leader,

Yasser Arafat, following two devastating suicide bombs in Jerusalem. In Cancun, Mexico, the World Trade Organisation is meeting, with the main debate focused on the regulation of agricultural trade. The outcome of the summit will have major ramifications for the regulation of the global capitalist economy and for the social and economic disparities between North and South. Outside are thousands of anti-globalisation protesters, drawn from around the world as part of a maturing global social movement and engaged in a game with police that involves the sophisticated use, control and manipulation of space. Meanwhile the population of Sweden is mourning the death of their Foreign Minister, murdered as she campaigned for a 'yes' vote in a referendum on membership of the European single currency – a vote that is regarded as a crucial decision for the project of European integration.

These are for the most part 'big p' political stories, to employ Flint's (2003) distinction which we referred to in Chapter 1, although it is notable that their effects inevitably seep into the 'small p' politics of everyday life. The stories of 'small p' politics are also happening all around us as we write, but rarely make the news headlines. Stories about the provision of public services, the use of public space, labour relations in the workplace, gender relations in the home, the meaning of landscapes and valued places, campaigns against proposed developments and for environmental protection, local politics and citizen participation, and so on.

We live in a dynamic world. The 'political geographies' that are our *objects of study* at the start of the twenty-first century are very different from those of twenty-five, fifty or 100 years ago. It is therefore not surprising to find that the academic discourse of political geography is also dynamic and responsive to the need to develop new concepts, draw on new theories, ask new questions and open up new avenues of enquiry. Ironically, however, political geography as an *academic discipline* has not always been quickest to react to the new agendas, and much of the most innovative recent work on politics and geography has been done by others – notably cultural geographers, social geographers and urban sociologists – a trend

that has for some commentators weakened 'political geography' as a discipline.

In Chapter 1 we noted that there is a debate about the future direction of political geography. One side argues that political geography needs to reassert its core concepts of territory, state and nation, to establish firm boundaries to its remit and to reclaim ground from cultural geography. The other side argues that the breadth, vitality and openness of contemporary political geography should be celebrated and extended. We have associated ourselves with the latter camp and in this book we have attempted to demonstrate the enormous potential of 'political geography' as it is practised today. As such, our examples have ranged from the political-economic analysis of the state to the cultural analysis of landscape and identity; from the local scale of building design and protest camps to the global scale of geopolitics; and from traditional concepts like territory and nationalism to newer theories and concepts such as the *régulation* approach and social capital.

That is not to say that we consider the traditional political geography concerns of territorial organisation, state power, geopolitics and electoral geographies to be unimportant – just that to concentrate on these themes alone would be to fail to adequately describe and interpret the emerging political geographies of the twenty-first century. At the same time, the geographical concepts that underlie these traditional concerns – ideas of space, place, scale and territory – remain essential to all areas of political geography and it is our sensitivity to the issues of how political processes interact with these geographical concepts that makes political geography distinct from other forms of political analysis. In this book we have drawn not only on work by geographers but also on work by political scientists, sociologists, economists, historians, urban theorists, cultural theorists and policy analysts. In some cases these writers have borrowed ideas from geography and applied them to their own work; in other cases it has been political geographers who have taken the original non-spatial research and translated it through a geographical filter – but what unites it all is the basic tenet that understanding geography is fundamental to understanding politics.

These exchanges with other disciplines have played a major role in shaping contemporary political geography and it is essential that the dialogue remains open. Indeed, there is much that could still be learnt in both directions. The relationship between geography and political science, for example, is perhaps one of the least explored cross-overs in the social sciences, particularly when it comes to the translation of political theory. While political geography is saturated with social and cultural theory, its use of political theory is strangely limited – yet, to take the theme of Chapter 8 as an example, the insights that could be gained from engaging with, say, theories of democracy and participation are considerable.

In this book we have been able to offer only a snapshot of political geography at a certain point in time. Even in the short period between our writing and you reading these words it is inevitable that the discourse of political geography will have been modified just a little bit further. New research on new topics will have been started, new findings presented, new ideas posited and new theories developed. Perhaps you too will contribute to this dynamism. We hope that we have demonstrated that political geography is all around us on an everyday basis, and as such we hope that we may have inspired you to explore some of the themes covered here further through your own project work, dissertations and theses.

Glossary

Active citizenship The idea that citizens are not the passive recipients of rights and state benefits, but have a responsibility to be actively involved in the governing process. Commonly associated with the strategy of governmentality (*q.v.*) of 'governing through communities'.

Capitalism A specific social and economic system that is divided into two classes: those owning the means of production (land, machinery and factories, etc.) and those selling labour power. Under the capitalist mode of production, labour power is exploited to provide surplus value (or profit) and capitalists compete for this profit through a system that necessitates the 'accumulation of capital' (see Box 4.1).

Capitalist world system See *World systems analysis*.

Citizenship A mark of belonging to a political entity or collective that both guarantees rights for the individual and carries responsibilities towards the collective. Citizenship codifies the relationship between the individual and the state (see Box 8.1).

Civic nationalism A type of nationalism that is based on the organisation of the state and which highlights the fact that nations are produced as a result of certain processes. It is often a more inclusive form of nationalism.

Colony A political and spatial form, often based on ideas of domination, which is created by the colonisation of one territory and people by a state, organisation or group of people. Thus the act of colonialism is characterised by unequal economic, political and cultural relationships.

Community A collective of individuals who share a mutual sense of identity and solidarity. Communities are frequently defined in terms of a territorial association, but need not necessarily be so.

Critical geopolitics A sub-field of political geography that critically analyses the production, circulation and consumption of geopolitical knowledge. (See also *Geopolitics*.)

Cultural turn The popularisation in human geography in the late 1980s and early 1990s of the study of cultural relations, processes and entities, including issues of identity, difference and representation. Associated with the use of qualitative research methods and the influence of cultural studies and of post-structuralist and postmodernist thought.

Democracy A political system based on the principle of government by the people through majority decision making.

Devolution A process whereby political power is transferred from a national state to regions within the state.

Discourse A body of knowledge that structures a particular way of understanding the world (see Box 1.4).

Electoral geography A sub-field of political geography that is concerned with the analysis of the spatial aspects of elections, including the influence of geographical factors on voting behaviour and election outcomes, and the spatial patterns of election results.

Elite A group or cluster of individuals exercising (or attributed with) a disproportionate concentration of power.

Elite theory A political theory that holds that power

is concentrated within an exclusive minority group in society.

Empire A political form which is based on a subservient relationship between a metropolitan state and other lands or people. The political form of empire is, thus, closely related to the process of imperialism.

Ethnic nationalism A type of nationalism that emphasises the common cultural and historical links between a named human population. It is often associated with ideas concerning the 'naturalness' of nations. It can, under certain circumstances, be linked with extreme and exclusionary forms of nationalism.

Federalism A political system which emphasises the notion of subsidiarity. In this context, it is believed that decisions should be made at the smallest practical spatial scale.

Feminism As a political movement, feminism advocates the right of women to equality in society; as an intellectual movement, feminism challenges masculinist discourses and approaches an understanding of the world from a female perspective.

Gentrification The renovation of property in relatively less favoured areas by and for affluent incomers, displacing lower-income groups.

Geopolitics A sub-field of political geography concerned with political relations between states, the external strategies of states and the global balance of power.

Geopolitik Literally the German translation of *geopolitics* (*q.v.*), but particularly associated with a partisan form of political geography practised in Germany in the 1920s and 1930s that was used to support the racist and expansionist policies of the Nazi party.

Gerrymandering The deliberate manipulation of the territory of electoral districts for partisan gain. Named after the nineteenth-century Governor of Massachusetts, Eldridge Gerry.

Globalisation The advanced interconnection and interdependence of localities across the world – economically, socially, politically and culturally.

Glocalisation A term coined by Erik Swyngedouw to emphasise the simultaneous erosion of power from the nation scale upwards to the global and downwards to the local.

Governmentality The techniques and strategies by which a society is rendered governable (see Box 2.2).

Heartland A geopolitical term referring to an area of central Eurasia, similar to the territory of present-day Russia, whose control, Mackinder argued, was crucial to the global balance of power. Also known as the 'pivot area'.

Identity politics The way in which people's politics, in recent years, are increasingly being shaped by aspects of their identity. This can be contrasted with the traditional dominance of class conflict within politics.

Ideology Ideas and arguments that often help to sustain a relationship of power and domination.

Landscape The assemblage of physical objects that comprise the visual surface appearance of an area of land.

Landscape of power The symbolic representation of power relations through the landscape (*q.v.*), including both monumental landscapes with an explicit political meaning and more subtle signifiers in everyday landscapes, such as the juxtapositioning of skyscrapers and slums.

Lebensraum Literally 'living space', a term Ratzel borrowed from biology to indicate the territory required for the comfortable existence of a state. The concept was used to justify expansionist policies.

Local state A collective term for the apparatuses of the state (*q.v.*) that exist and operate at a local scale, usually with reference to a specific locality (*q.v.*). The local state includes not only 'local government' (local-scale elected or appointed public authorities) but also the local branches of the judiciary and security agencies.

Locality A place defined at a local scale with a territorial expression. The term implies a spatial unit that can be attributed with distinct characteristics and differentiated from other localities (see Box 6.1).

Malapportionment A term in electoral geography (*q.v.*) referring to the disproportionate distribution of seats in a legislature to a geographical district compared with its entitlement on an objective population-based allocation.

Nation   A named human population that is perceived as possessing a common culture, customs and territory.

Nationalism   An ideology that seeks to promote the existence of nations within the world. In addition, it can refer to nations' attempts to reach the political goal of being constituted as a nation-state.

Nation-state   A political form in which the boundaries of a state and nation coincide.

Neighbourhood effect   A theory in electoral geography (*q.v.*) that suggests that voting behaviour is influenced by the geographical situation of the voter. In other words, residents of a neighbourhood are more likely to vote the same way than would be anticipated on the basis of social or economic characteristics.

Othering   A term, advocated by Edward Said, which refers to the act of emphasising the perceived weaknesses of marginalised groups as a way of stressing the alleged strength of those in positions of power.

Place   A point, or area, of space that can be identified through verbal, written, cartographic or visual representation.

Pluralism   A political theory that holds that power is widely dispersed within society and that a diverse range of groups have an equal opportunity to influence the political process.

Political economy   A wide range of different perspectives on the links between state, economy and society as sets of moving parts and founded on production, i.e. the social production of existence.

Post-structuralism   A philosophical movement of the late twentieth century that rejects notions of essential truth and the rational subject and proposes instead that meanings are produced within language and subjectivity is constructed through discourse (*q.v.*) (see Box 1.3).

Power   The capacity to do something. Politics can be described as the pursuit and discharge of power, but there are many different conceptualisations of exactly what power is and how it works (see Box 1.1).

Public policy   involves any form of deliberate intervention, regulation, governance, or prescriptive or alleviative action, by state or non-state bodies, intended to shape social, economic or environmental conditions (see Box 9.1).

Public space   An area of space to which in theory all people have a right of access without restriction, selection or payment. In practice, public space is regulated both formally and informally and the freedom to use public space differs between groups. Many public spaces are privately owned but open for public use by convention or for commercial purposes.

Quantitative revolution   A period of transformation in human geography in the 1950s and 1960s associated with the introduction of statistical and mathematical methods for geographical research and analysis, replacing the previous concern with areal differentiation and regional studies.

Region   A more or less bounded area possessing some relative unity or organising principle that distinguishes it from other regions. Regions, however, are never closed and are actively produced and reproduced by different forms of agency.

***Régulation* approach**   A set of neo-Marxist ideas on political economy, whereby economies and societies emerge through social, economic and institutional frameworks and supports, despite the instabilities and crisis tendencies within capitalism (see Box 4.2).

Resistance   The act of opposing or withstanding the exercise of power. Geographies of resistance are concerned with studying the spatial aspects of political opposition to the state and other centres of power.

Scale   A level of representation that is differentiated from other scales by variations in magnitude. Geographical scale is differentiated by the spatial dimensions encompassed by each level of magnitude and is commonly referred to in terms of fixed (but ambiguous) increments, including (with increasing magnitude): local, regional, national and global scale.

Social capital   The worth and potential that are invested in social networks and contacts between people (see Box 8.4).

Social construction   The ascribing of meaning to things by and through social interactions. A social construct has no fixed meaning outside the social context of its definition (see Box 6.3).

Social movement A network of individuals, groups and/or organisations engaged in political or cultural activity based on a shared identity. The components of a social movement may be highly fragmented and diverse in nature, with no single centre of leadership (see Box 8.7).

Spatial science A form of human geography that applied scientific principles and models to the analysis of spatial processes and spatial variations.

State A political organisation, possessing a degree of centrality, whose claim to authority is based on control of a defined territory.

Tactical voting The practice by which electors vote for their second preference candidate in order to prevent a third candidate from being elected.

Territory A space that is imbued with notions of power, domination and ownership.

Urban regime theory A model that proposes that stability in urban politics is achieved through the construction of 'urban regimes' that draw together the resources of public and private actors to produce a 'capacity to act'.

Workfare A model of welfare reform based on the movement from a universal rights and needs-based entitlement to income support and to a selective system combining welfare with work in order to enforce new social responsibilities. British discourses involve the term *welfare-to-work*, while in North America the term *work-first* is more common (see Box 4.9).

World systems analysis A conceptual approach that proposes that social change at any scale can be understood only in the context of a wider world system, and that change needs to be approached through a long-term historical perspective (see Box 1.2).

References

Abernethy, D.B. (2000) *The Dynamics of Global Dominance: European Overseas Empires*, 1415–1980, New Haven CT: Yale University Press.

Abu-Lughod, J. (1994) *From Urban Village to East Village: The Battle for New York's Lower East Side*, New York: Blackwell.

Aglietta, M. (1978) 'Phases of US capital expansion', *New Left Review*, 110: 17–28.

Aglietta, M. (2000) *A Theory of Capitalist Regulation: The US Experience*, new edition, London: Verso.

Agnew, J. (1995) 'The rhetoric of regionalism: the Northern League in Italian politics, 1983–94', *Transactions of the Institute of British Geographers*, 20: 156–72.

Agnew, J. (ed.) (1997) *Political Geography: A Reader*, London: Arnold; New York: Wiley.

Agnew, J. (1998) *Geopolitics*, London: Routledge.

Agnew, J. (2002a) *Making Political Geography*, London: Arnold.

Agnew, J. (2002b) *Place and Politics in Modern Italy*, Chicago: University of Chicago Press.

Agnew, J. (2003) 'Contemporary political geography: intellectual heterodoxy and its dilemmas', *Political Geography*, 22: 603–6.

Agnew, J. and Corbridge, S. (1995) *Mastering Space*, London: Routledge.

Agnew, J., Mitchell, K. and Toal, G. (eds) (2003) *A Companion to Political Geography*, Oxford: Blackwell.

Alexander, L. (1963) *World Political Patterns*, Chicago: Rand McNally.

Allen, J., Massey, D. and Cochrane, A. (1998) *Rethinking the Region*, London: Routledge.

Amin, A. (ed.) (1994) *Post-Fordism: A Reader*, Oxford: Blackwell.

Amin, A. and Robins, K. (1990) 'The re-emergence of regional economies? The mythical geography of flexible accumulation', *Environment and Planning D: Society and Space*, 8: 7–34.

Amin, A. and Thrift, N. (1994) 'Living in the global', in A. Amin and N. Thrift (eds) *Globalization, Institutions, and Regional Development in Europe*, Oxford: Oxford University Press.

Amin, A. and Thrift, N. (1997) 'Globalization, socio-economics, territoriality', in R. Lee and J. Wills (eds) *Geographies of Economies*, London: Arnold.

Anderson, B. (1983) *Imagined Communities: Reflections on the Origin and Spread of Nationalism*, London: Verso.

Anderson, J. (1988) 'Nationalist ideology and territory', in R.J. Johnston, D.G. Knight and E. Kaufman (eds) *Nationalism, Self-determination and Political Geography*, London: Croom Helm.

Anderson, J. (1995) 'The exaggerated death of the nation state', in J. Anderson, C. Brook and A. Cochrane (eds) *A Global World? Reordering Political Space*, Oxford: Oxford University Press.

Anderson, J. (1996) 'The shifting stage of politics: new medieval and postmodern territorialities', *Environment and Planning D: Society and Space*, 14: 133–53.

Archer, J.C. (2002) 'The geography of an interminable election: Bush *v.* Gore, 2000', *Political Geography*, 21: 71–7.

Area (1980) 'Observations. A future for cultural geography?' *Area*, 12: 105–13.

Arnold, G. (1989) *The Thirdworld Handbook*, London: Cassell.

Arrighi, G. and Silver, B.J. (2001) 'Capitalism and

world (dis)order', in M. Cox, T. Dunne and K. Booth (eds) *Empires, Systems, States: Great Transformations in International Politics*, Cambridge: Cambridge University Press.

Atkinson, D. (2000) 'Nomadic strategies and colonial governance: domination and resistance in Cyrenaica, 1923–1932', in J.P. Sharp, P. Routledge, C. Philo and R. Paddison (eds) *Entanglements of Power*, London: Routledge.

Atkinson, D. and Cosgrove, D. (1998) 'Urban rhetoric and embodied identities: city, nation and empire at the Vittorio Emanuel II monument in Rome, 1870–1945', *Annals of the Association of American Geographers*, 88: 28–49.

Atkinson, R. and Moon, G. (1994) *Urban Policy in Britain: The City, the State and the Market*, London: Macmillan.

Azaryahu, M. (1997) 'German reunification and the politics of street names: the case of East Berlin', *Political Geography*, 16: 479–94.

Azaryahu, M. and Kellerman, A. (1999) 'Symbolic places of national history and revival: a study in Zionist mythical geography', *Transactions of the Institute of British Geographers*, 24: 109–23.

Bachrach, P. and Baratz, M. (1962) 'Two faces of power', *American Political Science Review*, 56: 947–52.

Bachrach, P. and Baratz, M. (1970) *Power and Poverty: Theory and Practice*, Oxford: Oxford University Press.

Bakshi, P., Goodwin, M., Painter, J. and Southern, A. (1995) 'Gender, race and class in the local welfare state: moving beyond regulation theory in analysing the transition from Fordism', *Environment and Planning A*, 27: 1539–54.

Banks, M. and MacKian. S. (2000) 'Jump in! The water's warm. A comment on Peck's "grey geography"', *Transactions of the Institute of British Geographers*, 25: 249–54.

Baratz, M.S. and White, S.B. (1996) 'Childfare: a new direction for welfare reform', *Urban Geography*, 33: 1935–44.

Barnes, T. and Duncan, J. (eds) (1992) *Writing Worlds: Discourse, Text and Metaphor in the Representation of Landscape*, London: Routledge.

Bassett, K. and Short, K. (1980) *Housing and Residential Structure: Alternative Approaches*, London: Routledge & Kegan Paul.

Bassin, M. (1987) 'Imperialism and the nation state in Friedrich Ratzel's political geography', *Progress in Human Geography*, 11: 473–95.

Bayoumi, M. (2002) 'Letter to a G-man', in M. Sorkin and S. Zukin (eds) *After the World Trade Center*, New York: Routledge.

Bell, D. (1995) 'Pleasure and danger: the paradoxical spaces of sexual citizenship', *Political Geography*, 14: 139–53.

Bell, D. and Valentine, G. (eds) (1995) *Mapping Desire*, London: Routledge.

Bell, J.E. and Staeheli, L.A. (2001) 'Discourses of diffusion and democratization', *Political Geography*, 20: 175–95.

Belsey, C. (2002) *Poststructuralism: A Very Short Introduction*, Oxford: Oxford University Press.

Berry, B. (1969) Review of Russett, *International Regions and the International System*, *Geographical Review*, 59: 450.

Berry, B. (1972) 'More on relevance and policy analysis', *Area*, 4: 77–80.

Berry, B. (1994) 'Editorial. Let's have more policy analysis', *Urban Geography*, 15: 315–17.

Berry, J.M., Portney, K.E. and Thomson, K. (1993) *The Rebirth of Urban Democracy*, Washington DC: Brookings Institution.

Bery, L.D. and Kearns, R.A. (1996) 'Naming as norming: race, gender and identity politics of naming of places in Aotearoa, New Zealand', *Environment and Planning D: Society and Space*, 14: 99–122.

Billig, M. (1995) *Banal Nationalism*, London: Sage.

Blair, T. (2000) Speech on Britishness, London: Labour Party.

Blomley, N. (1994) 'Activism and the academy', *Environment and Planning D: Society and Space*, 12: 383–5.

Blowers, A.T. (1974) 'Relevance, research and the political process' *Area*, 6: 32–6.

Blunt, A., and Wills, J. (2000) *Dissident Geographies: An Introduction to Radical Ideas and Practice*, London: Pearson.

Boudreau, J-A. (2003) 'The politics of territorialization: regionalism, localism and other isms . . . the case of Montreal', *Journal of Urban Affairs*, 25: 179–99.

Bourdieu, P. (1984) *Distinction: A Social Critique of the Judgement of Taste*, Cambridge MA: Harvard University Press.

Bowman, I. (1921) *The New World: Problems in Political Geography*, New York: World Book Co.

Boyer, R. (1990) *The Regulation School: A Critical Introduction*, New York: Columbia University Press.

Boyer, R. and Saillard, Y. (eds) (2002) Régulation *Theory: The State of the Art*, London: Routledge.

Bradford Landau, S. and Condit, C. (1999) *The Rise of the New York Skyscraper*, New Haven CT: Yale University Press.

Braun, B. and Castree, N. (eds) (1998) *Remaking Reality: Nature and the Millennium*, London: Routledge.

Brenner, N. (2001) 'The limits to scale? Methodological reflections on scalar structuration', *Progress in Human Geography*, 25: 525–48.

Brenner, N. and Theodore, N. (eds) (2002) *Spaces of Neoliberalism: Urban Restructuring in North America and Western Europe*, Oxford: Blackwell.

Brenner, N. Jessop, B., Jones, M. and MacLeod, G. (2003) 'Introduction: state space in question', in N. Brenner, B. Jessop, M. Jones and G. MacLeod (eds) *State/Space: A Reader*, Oxford: Blackwell.

Brenner, R. and Glick, M. (1991) 'The regulation approach: theory and history', *New Left Review*, 188: 45–119.

Brodie, J. (1997) 'Meso-discourses, state forms, and the gendering of liberal-democratic citizenship', *Citizenship Studies*, 1: 223–42.

Brown, G. (1998) 'Address at the launch of the Tayside Pathfinder', London: HM Treasury.

Brown, M. (1997a) *RePlacing Citizenship: AIDS Activism and Radical Democracy*, New York: Guilford Press.

Brown, M. (1997b) 'Radical politics out of place?', in S. Pile and M. Keith (eds) *Geographies of Resistance*, London: Routledge.

Brusco, S. and Righi, E. (1989) 'Local government, industrial policy and social consensus: the case of Modena (Italy)', *Economy and Society*, 18: 405–24.

Buncombe, A. (2003) 'New York town decides to drop its Cromwell crest', *Independent*, 8 August.

Burch, M. and Wood, B. (1989) *Public Policy in Britain*, Oxford: Blackwell.

Burnett, A.D. and Taylor, P.J. (eds) (1981) *Political Studies from Spatial Perspectives*, Chichester: Wiley.

Busteed, M.A. (1975) *Geography and Voting Behaviour*, London: Oxford University Press.

Butler, D. and Stokes, D.E. (1969) *Political Change in Britain*, London: Macmillan.

Byrne, T. (2000) *Local Government in Britain*, London: Penguin.

Calvocoressi, P. (1991) *World Politics since 1945*, London: Longman.

Castells, M. (1977) *The Urban Question: A Marxist Approach*, London: Arnold.

Castells, M. (1978) *City, Class and Power*, London: Macmillan.

Castells, M. (1983) *The City and the Grassroots*, Berkeley CA: University of California Press.

Castree, N. (1999a) 'Envisioning capitalism: geography and the renewal of Marxian political economy', *Transactions of the Institute of British Geographers*, 24: 137–58.

Castree, N. (1999b) 'Out there? In here? Domesticating critical geography', *Area*, 21: 81–6.

Cerny, C. (1997) 'Paradoxes of the competition state: the dynamics of political globalization', *Government and Opposition*, 32: 251–74.

Christophers, B. (1998) *Positioning the Missionary*, Vancouver: University of British Columbia Press.

Clapham, D., Kintrea, K. and Kay, H. (2000) 'User participation in community housing: is small really beautiful?' in G. Stoker (ed.) *The New Politics of British Local Governance*, London: Macmillan.

Clark, G. and Dear, M. (1984) *State Apparatus*, London: Allen & Unwin.

Clarke, D.B. and Doel, M.A. (1994) 'Transpolitical geography', *Geoforum*, 25: 505–24.

Clarke, S. and Gaile, G. (1998) *The Work of Cities*, Minneapolis MN: University of Minnesota Press.

Clayton, D. (2000) 'Imperialism', in R.J. Johnston, D. Gregory, G. Pratt and M. Watts (eds) *The Dictionary of Human Geography*, Oxford: Blackwell.

Clegg, S.R. (1989) *Frameworks of Power*, London: Sage.

Cloke, P., Philo, C. and Sadler, D. (1991) *Approaching Human Geography*, London: Chapman.

Cochrane, A., Peck, J. and Tickell, A. (1996) 'Manchester plays games: exploring the local politics of globalisation', *Urban Studies*, 33: 1319–36.

Cockburn, C. (1977) *The Local State: Management of Cities and People*, London: Pluto Press.

Cohen, A. (1980) 'Drama and politics in the development of a London carnival', *Man*, 15: 65–87.

Cohen, S.B. and Rosenthal, L.D. (1971) 'A geographical model for political systems analysis', *Geographical Review*, 61: 5–31.

Cole, J.P. (1959) *Geography of World Affairs*, Harmondsworth: Penguin.

Collard, S. (1996) 'Politics, culture and urban transformation in Jacque Chirac's Paris, 1977–1995', *French Cultural Studies*, 7: 1–31.

Conversi, D. (1995) 'Reassessing current theories of nationalism: nationalism as boundary maintenance and creation', in J. Agnew (ed.) *Political Geography: A Reader*, London: Arnold. Reprinted from *Nationalism and Ethnic Politics*, 1: 73–85.

Cooke, P. (ed.) (1989) *Localities*, London: Unwin Hyman.

Coppock, J.T. (1974) 'Geography and public policy: challenges, opportunities and implications', *Transactions of the Institute of British Geographers*, 63: 1–16.

Cornog, E.W. (1988) '"To give character to our city": New York's City Hall', *New York History*, 69: 389–423.

Cosgrove, D. and Daniels, S. (1988) *The Iconography of Landscape*, Cambridge: Cambridge University Press.

Cox, K.R. (1998) 'Spaces of dependence, spaces of engagement and the politics of scale, or: Looking for local politics', *Political Geography*, 17: 1–24.

Cox, K. (2002) *Political Geography: Territory, State and Society*, Oxford: Blackwell.

Cox, K. (2003) 'Political geography and the territorial', *Political Geography*, 22: 607–10.

Cox, K. and Johnston, R.J. (eds) (1982) *Conflict, Politics and the Urban Scene*, London: Longman.

Cox, K. and Mair, A. (1988) 'Locality and community in the politics of local economic development', *Annals of the Association of American Geographers*, 78: 307–25.

Cox, K. and Mair, A. (1991) 'From localised social structures to localities as agents', *Environment and Planning A*, 23: 197–213.

Crang, M. (1999) 'Nation, region and homeland: history and tradition in Dalarna', Sweden, *Ecumene*, 6: 447–70.

Crang, P. (1999) 'Local–global', in P. Cloke, P. Crang and M. Goodwin (eds) *Introducing Human Geographies*, London: Arnold.

Dahl, R. (1957) 'The concept of power', *Behavioural Science*, 2: 201–5.

Dahl, R. (1958) 'Critique of the ruling elite model', *American Political Science Review*, 52: 463–9.

Dahl, R. (1961) *Who Governs? Democracy and Power in an American City*, New Haven CT: Yale University Press.

Daniels, S. (1993) *Fields of Vision: Landscape Imagery and National Identity in England and the United States*, Cambridge: Polity Press.

Dean, M. (1999) *Governmentality: Power and Rule in Modern Society*, London: Sage.

Dear, M. (1999) 'The relevance of postmodernism', *Scottish Geographical Journal*, 115: 143–50.

De Bernieres, L. (1991) *The War of Don Emmanuel's Nether Parts*, London: Minerva.

De Bernieres, L. (1992) *Señor Vivo and the Coca Lord*, London: Minerva.

De Bernieres, L. (1993) *The Troublesome Offspring of Cardinal Guzman*, London: Minerva.

De Bilj, H.J. (1967) *Systematic Political Geography*, New York: Wiley.

de Burca, G. and Scott, J. (eds) (2001) *The EU and the WTO: Legal and Constitutional Aspects*, Oxford: Hart.

Delaney, D. and Leitner, H. (1997) 'The political construction of scale', *Political Geography*, 16: 93–7.

Deleuze, G. and Guattari, F. (1988) *A Thousand Plateaus: Capitalism and Schizophrenia*, trans. B. Massumi, London: Athlone Press.

della Porta, D. and Diani, M. (1999) *Social Movements: An Introduction*, Oxford: Blackwell.

Diani, M. (1992) 'The concept of social movement', *Sociological Review*, 40: 1–25.

Dicken, P. (1998) *Global Shift: Transforming the World Economy*, London: Chapman.

Dicken, P. (2003) *Global Shift: Reshaping the Global Economic Map in the Twenty-first Century*, London: Sage.

Dicken, P., Peck, J. and Tickell, A. (1997) 'Unpacking globalisation', in R. Lee and J. Wills (eds) *Geographies of Economies*, London: Arnold.

Dickinson, R.E. (1976) *Regional Concept: The Anglo-American Leaders*, London: Routledge & Kegan Paul.

Dikshit, R.D. (1977) 'The retreat from political geography', *Area*, 9: 234–9.

DiGiovanna, S. (1996) 'Industrial districts and regional economic development: a regulation approach', *Regional Studies*, 30: 373–86.

Dodds, K. (1994) 'Geopolitics in the Foreign Office: British representations of Argentina, 1945–1961', *Transactions of the Institute of British Geographers*, 19: 273–90.

Dodds, K. (2000) *Geopolitics in a Changing World*, London: Prentice Hall.

Donnan, H. and Wilson, T.W. (1999) *Borders: Frontiers of Identity, Nation and State*, Oxford: Berg.

Dorling, D. and Shaw, M. (2002) 'Geographies of the agenda: public policy, the discipline and its (re)"turns"', *Progress in Human Geography*, 26: 629–46.

Drake, C. and Horton, J. (1983) 'Comment on editorial essay: sexist bias in political geography', *Political Geography Quarterly*, 2: 329–37.

Duncan, J. (1992) *The City as Text*, Cambridge: Cambridge University Press.

Duncan, J. (1993) 'Representing power: the politics and poetics of urban form in the Kandyan kingdom', in J. Duncan and D. Ley (eds) *Place/Culture/Representation*, London: Routledge.

Duncan, S. (1974) 'Research directions in social geography: housing opportunities and constraints', *Transactions of the Institute of British Geographers*, New Series, 1: 10–19.

Duncan, S. (1989) 'What is a locality?', in R. Peet and N. Thrift (eds) *New Models in Geography* II, London: Unwin Hyman.

Duncan, S. and Goodwin, M. (1988) *The Local State and Uneven Development: Behind the Local Government Crisis*, Cambridge: Polity Press.

Duncan, S. and Savage, M. (1989) 'Space, scale and locality', *Antipode*, 21: 179–206.

Duncan, S. and Savage, M. (1991) 'Commentary', *Environment and Planning A*, 23: 155–64.

Dunleavy, P. (1979) 'The urban basis of political alignment: social class, domestic property ownership and state intervention in consumption practices', *British Journal of Political Science*, 9: 409–43.

Dunleavy, P. (1980) *Urban Political Analysis: The Politics of Collective Consumption*, London: Macmillan.

Dunn, J. (1995) 'Introduction: crisis of the nation state?', in J. Dunn (ed.) *Contemporary Crisis of the Nation State?* Oxford: Blackwell.

East, W.G. (1937) 'The nature of political geography', *Politica*, 2: 259–86.

East, W.G. and Moodie, A.E. (1956) *The Changing World: Studies in Political Geography*, London: Harrap.

Edensor, T. (1997) 'National identity and the politics of memory: remembering Bruce and Wallace in symbolic space', *Environment and Planning D: Society and Space*, 29: 175–94.

Eder, K. (1993) *The New Politics of Class: Social Movements and Cultural Dynamics in Advanced Societies*, London: Sage.

Ehrenhalt, A. (1991) *The United States of Ambition*, New York: Times Books.

England, K. (2003) 'Towards a feminist political geography?' *Political Geography*, 22: 611–16.

England, K. and Stiell, B. (1997) ' "They think you're as stupid as your English is": constructing foreign domestic workers in Toronto', *Environment and Planning A*, 29: 195–215.

Esping-Andersen, G. (1990) *The Three Worlds of Welfare Capitalism*, Cambridge: Polity Press.

Esser, J. and Hirsch, J. (1989) 'The crisis of Fordism and the dimensions of a "postfordist" regional and urban structure', *International Journal of Urban and Regional Research*, 13: 417–37.

Etherington, D. and Jones, M. (2004) 'Beyond contradictions of the workfare state? Denmark, welfare-*through*-work, and the promise of job rotation', *Environment and Planning C: Government and Policy*, 22, 129–48.

Etzioni-Halevy, E. (1993) *The Elite Connection*, Cambridge: Polity Press.

Fainstein, S.S. and Hirst, C. (1995) 'Urban social movements', in D. Judge, G. Stoker and H. Wolman (eds) *Theories of Urban Politics*, London: Sage.

Fevre, R., Borland, J. and Denney, D. (1999) 'Nation, community and conflict: housing policy and immigration in North Wales', in R. Fevre and A. Thompson (eds) *Nation, Identity and Social Theory: Perspectives from Wales*, Cardiff: University of Wales Press.

Fischer, C.S., Jackson, R.M., Stueve, C.A., Gerson, K., Jones, L.M. and Baldassare, M. (1977) *Networks and Places*, New York: Free Press.

Flint, C. (2003) 'Dying for a "P"? Some questions facing contemporary political geography', *Political Geography*, 22: 617–20.

Florida, R. and Jonas, A. (1991) 'US urban policy: the postwar state and capitalist regulation', *Antipode*, 23: 349–84.

Foucault, M. (1973) [1966] *The Order of Things: An Archaeology of the Human Sciences*, New York: Vintage Books.

Foucault, M. (1974) [1969] *The Archaeology of Knowledge*, London: Tavistock.

Foucault, M. (1977) *Discipline and Punish: The Birth of the Prison*, London: Allen Lane.

Foucault, M. (1979) *The History of Sexuality* I, *An Introduction*, London: Allen Lane.

Foucault, M. (1980) *Power/Knowledge*, trans. C. Gordon, London: Harvester Wheatsheaf.

Foucault, M. (1984) 'Space, Knowledge and Power', in P. Rabinow (ed.) *The Foucault Reader*, London: Penguin, pp. 239–56.

Foucault, M. (1991) 'Governmentality', in G. Burchell, C. Gordon and P. Miller (eds) *The Foucault Effect: Studies in Governmentality*, London: Harvester Wheatsheaf.

Fyfe, N.R. and Bannister, J. (1996) 'City watching: closed circuit television surveillance in public spaces', *Area*, 28: 37–46.

Gamble, A. (1988) *The Free Economy and the Strong State: The Politics of Thatcherism*, London: Macmillan.

Gellner, E. (1983) *Nations and Nationalism*, Oxford: Blackwell.

Gerth, H. and Mills, C.W. (1970) *From Max Weber: Essays in Sociology*, London: Routledge & Kegan Paul.

Gibson-Graham, J.K. (1996) *The End of Capitalism (as we knew it): A Feminist Critique of Political Economy*, Oxford: Blackwell.

Giddens, A. (1985) *The Nation State and Violence*, Cambridge: Polity Press.

Gilbert, E. and Helleiner, E. (eds) (1999) *Nation States and Money: The Past, Present and Future of National Currencies*, London: Routledge.

Giordano, B. (2001) ' "Institutional thickness" and the resurgence of (the "new") regionalism in Italy: a case study of the Northern League in the province of Varese', *Transactions of the Institute of British Geographers*, 25: 25–42.

Gittell, R. and Vidal, A. (1998) *Community Organizing: Building Social Capital as a Development Strategy*, Thousand Oaks CA: Sage.

Glassner, M. (1996) *Political Geography*, New York: Wiley.

Goblet, Y. (1955) *Political Geography and the World Map*, New York: Praeger.

Godlewska, A. and Smith, N. (1994) *Geography and Empire*, Oxford: Blackwell.

Goffman, E. (1971) [1959] *The Presentation of Self in Everyday Life*, London: Penguin.

Goodwin, M. (1992) 'The changing local state', in P. Cloke (ed.) *Policy and Change in Thatcher's Britain*, Oxford: Pergamon Press.

Goodwin, M. (2001) 'Regulation as process: regulation theory and comparative urban and regional research', *Journal of Housing and the Built Environment*, 16: 71–87.

Goodwin, M. and Painter, J. (1996) 'Local governance, the crises of Fordism and the changing geographies of regulation', *Transactions of the Institute of British Geographers*, 21: 635–48.

Goodwin, M., Cloke, P. and Milbourne, P. (1995) 'Regulation theory and rural research: theorising contemporary rural change', *Environment and Planning A*, 27: 1245–60.

Goodwin, M., Duncan, S. and Halford, S. (1993) 'Regulation theory, the local state and the transition in urban politics', *Environment and Planning D: Society and Space*, 11: 67–88.

Goodwin, M., Johnstone, C. and Williams, K. (2000) 'CCTV surveillance in urban Britain: beyond the rhetoric of crime prevention', in J. Gold and G. Revill (eds) *Landscapes of Defence*, London: Prentice Hall.

Gordon, C. (1991) 'Introduction', in G. Burchell, C. Gordon and P. Miller (eds) *The Foucault Effect: Studies in Governmentality*, London: Harvester Wheatsheaf.

Graham, J. (1992) 'Post-Fordism as politics: the political consequences of narratives on the left', *Environment and Planning D: Society and Space*, 10: 393–410.

Gramsci, A. (1971) *Selections from the Prison Notebooks of Antonio Gramsci*, London: Lawrence & Wishart.

Gregory, D. (2000a) 'Discourse', in R. Johnston, D. Gregory, G. Pratt and M. Watts (eds) *The Dictionary of Human Geography*, Oxford: Blackwell.

Gregory, D. (2000b) 'Positivism', in R. Johnston, D. Gregory, G. Pratt and M. Watts (eds) *The Dictionary of Human Geography*, Oxford: Blackwell.

Gregson, N. (1987) 'The CURS initiative: some further comments', *Antipode*, 19, 364–70.

Gruffudd, P. (1994) 'Back to the land: historiography, rurality and the nation in interwar Wales', *Transactions of the Institute of British Geographers*, 19: 61–77.

Halfacree, K. (1994) 'The importance of "the rural" in the constitution of counterurbanization: evidence from England in the 1980s', *Sociologia Ruralis*, 34, 164–89.

Hall, P. (1974) 'The new political geography', *Transactions of the Institute of British Geographers*, 63: 48–52.

Hall, P. (1987) *Urban and Regional Planning*, London: Allen & Unwin.

Ham, C. and Hill, C. (1993) *The Policy Process in the Modern Capitalist State*, London: Harvester Wheatsheaf.

Harding, A. (1995) 'Elite theory and growth machines', in D. Judge, G. Stoker and H. Wolman (eds) *Theories of Urban Politics*, London: Sage.

Hartshorne, R. (1950) 'The functional approach in political geography', *Annals of the Association of American Geographers*, 40: 95–130.

Hartshorne, R. (1954) 'Political geography', in P.E. James and C.F. Jones (eds) *American Geography: Inventory and Prospects*, Syracuse NY: Syracuse University Press.

Harvey, D. (1973) *Social Justice and the City*, Oxford: Blackwell.

Harvey, D. (1974) 'What kind of geography for what kind of public policy?' *Transactions of the Institute of British Geographers*, 63: 18–24.

Harvey, D. (1979) 'Monument and myth', *Annals of the Association of American Geographers*, 69: 362–81.

Harvey, D. (1989a) *The Condition of Postmodernity*, Oxford: Blackwell.

Harvey, D. (1989b) 'From managerialism to entrepreneurialism: the transformation in urban governance in late capitalism', *Geografiska Annaler*, 71B: 3–17.

Harvey, D. (2003) *The New Imperialism*, Oxford: Oxford University Press.

Hawthorn, G. (1995) 'The crises of southern states', in J. Dunn (ed.) *Contemporary Crisis of the Nation State?* Oxford: Blackwell.

Hay, C. (1995) 'Re-stating the problem of regulation and re-regulating the local state', *Economy and Society*, 24: 287–307.

Hayden, D. (1995) *The Power of Place: Urban Landscape as Public History*, Cambridge MA: MIT Press.

Heffernan, M. (1995) 'For ever England: the western front and the politics of remembrance in Britain', *Ecumene*, 2: 293–324.

Held, D., McGrew, A., Goldblatt, D. and Perraton, J. (1999) *Global Transformations: Politics, Economics and Culture*, Cambridge: Polity Press.

Herb, G.H. (1997) *Under the Map of Germany: Nationalism and Propaganda, 1918–1945*, London: Routledge.

Herbert, S. (1996) 'The geopolitics of the police: Foucault, disciplinary power and the tactics of the Los Angeles Police Department', *Political Geography*, 15: 47–61.

Herbert, S. (1997) *Policing Space: Territoriality and the Los Angeles Police Department*, Minneapolis MN: University of Minnesota Press.

Herbert-Cheshire, L. (2000) 'Contemporary strategies for rural community development in Australia: a governmentality perspective', *Journal of Rural Studies*, 16: 203–15.

Herod, A. (1995) 'International labor solidarity and the geography of the global economy', *Economic Geography*, 71: 341–63.

Herod, A. (1997) 'Labor's spatial praxis and the geography of contract bargaining in the US east coast longshore industry, 1953–1989', *Political Geography*, 16: 145–70.

Heske, H. (1987) 'Karl Haushofer: his role in German geopolitics in the Nazi period', *Political Geography Quarterly*, 5: 267–82.

Hirst, P. and Thompson, G. (1996) *Globalization in Question: The International Economy and the Possibilities of Governance*, Cambridge: Polity Press.

Hoggart, K. (1996) 'All washed up and nowhere to go? Public policy and geographical research', *Progress in Human Geography*, 20: 110–22.

Hoggart, K. (1997) 'Uncertainty washed with conviction in the investigation of public policy', *Progress in Human Geography*, 21: 109–18.

Holmes, L. and Ettinger, S. (1997) *Workfairness and the Struggle for Jobs, Justice and Equality*, New York: Workfairness.

Hooson, I.D. (ed.) (1994) *Geography and National Identity*, Oxford: Blackwell.

House, J.W. (1973) 'Geographers, decision takers and policy matters', in M. Chisholm and B. Rodgers (eds) *Studies in Human Geography*, London: Heinemann/SSRC.

Huber, E. and Stephens, J.D. (2001) *Development and Crisis of the Welfare State*, Chicago: University of Chicago Press.

Hudson, R. and Williams, A. (1995) *Divided Britain*, Chichester: Wiley.

Hunter, F. (1953) *Community Power Structure*, Chapel Hill NC: University of North Carolina Press.

Hunter, F. (1980) *Community Power Succession: Atlanta's Policy Makers Revisited*, Chapel Hill NC: University of North Carolina Press.

Huntington, S.P. (1991) *Third Wave: Democratization in the Late Twentieth Century*, Norman OK: University of Oklahoma Press.

Hyndman, J. (2001) 'Towards a feminist geopolitics', *Canadian Geographer*, 45: 210–22.

Ikenberry, G.J. (2001) 'American power and the empire of capitalist democracy', in M. Cox, T. Dunne and K. Booth (eds) *Empires, Systems, States: Great Transformations in International Politics*, Cambridge: Cambridge University Press.

Isin, E. (2002) *Being Political: Genealogies of Citizenship*, Minneapolis MN: University of Minnesota Press.

Isin, E. and Turner, B.S. (eds) (2002) *The Handbook of Citizenship Studies*, London: Sage.

Jackson, P. (1988) 'Street life: the politics of carnival', *Society and Space*, 6: 213–27.

Jackson, P. (1989) *Maps of Meaning: An Introduction to Cultural Geography*, London: Unwin Hyman.

Jackson, P. (1992) 'The politics of the street: a geography of Caribana', *Political Geography*, 11: 130–51.

Jenson, J. (1991) 'All the world's a stage: ideas, space and time in Canadian political economy', *Studies in Political Economy*, 36: 43–72.

Jessop, B. (1990a) *State Theory: Putting the Capitalist State in its Place*, Cambridge: Polity Press.

Jessop, B. (1990b) 'Regulation theories in retrospect and prospect', *Economy and Society*, 19: 153–216.

Jessop, B. (1992) 'Fordism and post-Fordism: a critical reformulation', in M. Storper and A. Scott (eds) *Pathways to Industrialization and Regional Development*, London: Routledge.

Jessop, B. (1994) 'Post-Fordism and the state', in A. Amin (ed.) *Post-Fordism: A Reader*, Oxford: Blackwell.

Jessop, B. (1995) 'Towards a Schumpeterian workfare regime in Britain? Reflections on regulation, governance and welfare state', *Environment and Planning A*, 27: 1613–26.

Jessop, B. (1997a) 'Twenty years of the (Parisian) regulation approach: the paradox of success and failure at home and abroad', *New Political Economy*, 2: 503–26.

Jessop, B. (1997b) Survey article 'The regulation approach', *Journal of Political Philosophy*, 3: 287–326.

Jessop, B. (1997c) 'A neo-Gramscian approach to urban

regimes', in M. Lauria (ed.) *Reconstructing Urban Regime Theory: Regulating Urban Politics in a Global Economy*, London: Sage.

Jessop, B. (2001) 'Institutional (re)turns and the strategic-relational approach', *Environment and Planning A*, 33: 1213–35.

Jessop, B. (2002) *The Future of the Capitalist State*, Cambridge: Polity.

Jessop, B., Bonnett, K., Bromley, S. and Ling, T. (1988) *Thatcherism: A Tale of Two Nations*, Cambridge: Polity Press.

John, P. (2001) *Local Governance in Western Europe*, London: Sage.

Johnson, N. (1995) 'Cast in stone: monuments, geography and nationalism', *Environment and Planning D: Society and Space*, 13: 51–66.

Johnson, N. (1997) 'Making space: Gaeltacht policy and the politics of identity', in B. Graham (ed.) *In Search of Ireland: A Cultural Geography*, London: Routledge.

Johnston, R.J. (1979) *Political, Electoral and Spatial Systems*, Oxford: Oxford University Press.

Johnston, R.J. (1980) 'Political geography without politics', *Progress in Human Geography*, 4: 439–46.

Johnston, R.J. (1981) 'British political geography since Mackinder: a critical review', in A.D. Burnett and P.J. Taylor (eds) *Political Studies from Spatial Perspectives*, Chichester: Wiley.

Johnston, R.J. (1982) *Geography and the State: An Essay in Political Geography*, London: Macmillan.

Johnston, R.J. (1989) 'The state, political geography, geography', in R. Peet and N. Thrift (eds) *New Models in Geography* I, London: Unwin Hyman.

Johnston, R.J. (2000) 'Public policy, Geography and', in R.J. Johnston, D. Gregory, G. Pratt and M. Watts (eds) *The Dictionary of Human Geography*, Oxford: Blackwell.

Johnston, R.J. (2002a) 'Manipulating maps and winning elections: measuring the impact of mal-apportionment and gerrymandering', *Political Geography*, 21: 1–31.

Johnston, R.J. (2002b) 'If it isn't a gerrymander, what is it?' *Political Geography*, 21: 55–65.

Johnston, R.J. and Pattie, C. (1997) 'Where's the difference? Decomposing the impact of local election campaigns in Great Britain', *Electoral Studies*, 16: 165–74.

Johnston, R.J., Pattie, C. and Allsopp, G. (1988) *A Nation Dividing? The Electoral Map of Great Britain, 1979–1987*, Harlow: Longman.

Johnston, R.J., Pattie, C. and Rossiter, D. (2001) 'He lost . . . but he won! Electoral bias and George W. Bush's victory in the US presidential election, 2000', *Representation*, 38: 150–8.

Johnston, R.J., Rossiter, D. and Pattie, C. (1999) *The Boundary Commissions: Redrawing the UK's Map of Parliamentary Constituencies*, Manchester: Manchester University Press.

Johnston, R.J., Gregory, D., Pratt, G., and Watts, M. (eds) (2000) *The Dictionary of Human Geography*, Oxford: Blackwell.

Jones, C. (1996) 'Empire', in I. McLean (ed.) *Oxford Concise Dictionary of Politics*, Oxford: Oxford University Press.

Jones, M. (1996) 'Full steam ahead to a workfare state? Analysing the UK Employment Department's abolition', *Policy and Politics*, 24, 137–57.

Jones, M. (1997) 'Spatial selectivity of the state? The regulationist enigma and local struggles over economic governance', *Environment and Planning A*, 29: 831–64.

Jones, M. (1999) *New Institutional Spaces: Training and Enterprise Councils and the Remaking of Economic Governance*, London: Routledge/Regional Studies Association.

Jones, M. and Jones, R. (forthcoming) 'Nation states, ideological power and globalisation: can geographers catch the boat?', *Geoforum*.

Jones, M. and Ward, K. (2002) 'Excavating the logic of British urban policy: neoliberalism as the "crisis of crisis management"', *Antipode*, 34: 479–500.

Jones, M. and Ward, K. (forthcoming) 'Capitalist development and crisis theory: towards a "fourth-cut"?' *Antipode*.

Jones, R. and Desforges, L. (2003) 'Localities and the reproduction of Welsh nationalism', *Political Geography*, 22: 271–92.

Joseph, M. (2002) *Against the Romance of Community*, Minneapolis MN: University of Minnesota Press.

Judge, D., Stoker, G. and Wolman, H. (eds) (1995) *Theories of Urban Politics*, London: Sage.

Kasperson, R. and Minghi, J. (eds) (1969) *The Structure of Political Geography*, Chicago: Aldine.

Kearns, A. (1992) 'Active citizenship and urban governance', *Transactions of the Institute of British Geographers*, 17: 20–34.

Kearns, A. (1995) 'Active citizenship and local governance: political and geographical dimensions', *Political Geography*, 14: 155–75.

Kedourie, E. (1960) *Nationalism*, London: Hutchinson.

Kedourie, E. (ed.) (1971) *Nationalism in Asia and Africa*, London: Weidenfeld & Nicolson.

Kettle, M. (2001) 'Fate of flag stirs Mississippi voters to have their say', *Guardian* (London), 18 April, 13.

King, D. (1995) *Actively Seeking Work? The Politics of Unemployment and Welfare Policy in the United States and Great Britain*, Chicago: The University of Chicago Press.

King, R. (1995) 'Migrations, globalization and place', in D. Massey and P. Jess (eds) *A Place in the World? Places, Cultures and Globalization*, Oxford: Open University/Oxford University Press.

Kirby, A. (1993) *Power/Resistance: Local Politics and the Chaotic State*, Bloomington IN and Indianapolis IN: Indiana University Press.

Kofman, E. (1995) 'Citizenship for some but not for others: spaces of citizenship in contemporary Europe', *Political Geography*, 14: 121–37.

Kofman, E. (2002) 'Contemporary European migration, civic stratification and citizenship', *Political Geography*, 21: 1035–54.

Kofman, E. (2003) 'Future directions in political geography', *Political Geography*, 22, 621–24.

Kofman, E. and Peake, L. (1990) 'Into the 1990s: a gendered agenda for political geography', *Political Geography Quarterly*, 9: 313–36.

Koht, H. (1947) 'The dawn of nationalism in Europe', *American Historical Review*, 52: 265–80.

Koskela, H. (2002) 'Video surveillance, gender and the safety of public urban spaces: 'Peeping Tom' goes high-tech?' *Urban Geography*, 23: 257–78.

Krasner, S.D. (2001) 'Rethinking the sovereign state model', in M. Cox, T. Dunne and K. Booth (eds) *Empires, Systems, States: Great Transformations in International Politics*, Cambridge: Cambridge University Press.

Krätke, S. (1999) 'A regulationist approach to regional studies', *Environment and Planning A*, 31: 683–704.

Kula, W. (1986) *Measures and Men*, trans. R. Szreter, Princeton NJ: Princeton University Press.

Larner, W. (2000) 'Neoliberalism: policy, ideology, governmentality', *Studies in Political Economy*, 63: 5–26.

Latour, B. (1986) 'The powers of association', in J. Law (ed.) *Power, Action and Belief: A New Sociology of Knowledge*, London: Routledge.

Lauria, M. (ed.) (1997) *Reconstructing Urban Regime Theory: Regulating Urban Politics in a Global Economy*, London: Sage.

Leach, B. (1974) 'Race, problems and geography', *Transactions of the Institute of British Geographers*, 63: 41–7.

Leadbeater, C. (2000) *Living on Thin Air: The New Economy*, Harmondsworth: Penguin.

Lechte, J. (1994) *Fifty Key Contemporary Thinkers*, London and New York: Routledge.

Lee, R. (1995) 'Look after the pounds and the people will look after themselves: social reproduction, regulation, and social exclusion in Western Europe', *Environment and Planning A*, 27: 1577–94.

Leitner, H. (1990) 'Cities in pursuit of economic growth', *Political Geography Quarterly*, 9: 146–70.

Lenin, V.I. (1996) [1917] *Imperialism: The Highest Stage of Capitalism*, ed. N. Lewis and J. Malone, London: Junius.

Levi, M. (1996) 'Social and unsocial capital: a review essay of Robert Putnam's *Making Democracy Work*', *Politics and Society*, 24: 45–55.

Ley, D. and Olds, K. (1988) 'Landscape as spectacle: world's fairs and the culture of heroic consumption', *Environment and Planning D: Society and Space*, 6: 191–212.

Liepins, R. (2000) 'New energies for an old idea: reworking approaches to "community" in contemporary rural studies', *Journal of Rural Studies*, 16, 23–35.

Lindblom, C.E. (1968) *The Policy-making Process*, Englewood Cliffs NJ: Prentice Hall.

Lipietz, A. (1988) 'Reflections on a tale: the Marxist

foundations of the concepts of accumulation and regulation', *Studies in Political Economy*, 26: 7–36.

Lødemel, I. and Trickey, H. (eds) (2000) *An Offer you can't Refuse: Workfare in International Perspective*, Bristol: Policy Press.

Logan, J. and Molotch, H. (1987) *Urban Futures: The Political Economy of Place*, Berkeley CA: University of California Press.

Lovering, J. (1990) 'Fordism's unknown successor: a comment on Scott's theory of flexible accumulation', *International Journal of Urban and Regional Research*, 14: 159–74.

Low, M. (1997) 'Representation unbound', in K. Cox (ed.) *Spaces of Globalization: Reasserting the Power of the Local*, New York: Guilford Press.

Low, M. (2003) 'Political geography in question', *Political Geography*, 22: 625–31.

Lowndes, V. (1995) 'Citizenship and urban politics', in D. Judge, G. Stoker and H. Wolman (eds) *Theories of Urban Politics*, London: Sage.

Lukes, S. (1974) *Power: A Radical View*, London: Macmillan.

Lukes, S. (ed.) (1986) *Power*, Oxford: Blackwell.

McAllister, I. (1997) 'Regional voting', *Parliamentary Affairs*, 50: 641–57.

McCann, E. (2003) 'Framing space and time in the city: urban policy and the politics of spatial and temporal scale', *Journal of Urban Affairs*, 25: 159–78.

McGreal, C. (2001) 'An army guarding power and profits', *Guardian*, 29 May.

McKay, G. (ed.) (1998) *DiY Culture: Party and Protest in Nineties Britain*, London: Verso.

Mackinder, H.J. (1902) *Britain and the British Seas*, Oxford: Clarendon Press.

Mackinder, H.J. (1904) 'The geographical pivot of history', *Geographical Journal*, 23: 421–44.

Mackinder, H.J. (1919) *Democratic Ideals and Reality*, London: Constable.

MacLeod, G. (1997) 'Globalising Parisian thought waves: recent advances in the study of social regulation, politics, discourse and space', *Progress in Human Geography*, 21: 530–53.

MacLeod, G. (2001) 'New regionalism reconsidered: globalization and the remaking of political economic space', *International Journal of Urban and Regional Research*, 25: 804–29.

MacLeod, G. and Goodwin, M. (1999a) 'Reconstructing an urban and regional political economy: on the state, politics, scale, and explanation', *Political Geography*, 18: 697–730.

MacLeod, G. and Goodwin, M. (1999b) 'Space, scale and state strategy: rethinking urban and regional governance', *Progress in Human Geography*, 23: 503–27.

McPhail, I.R. (1971) 'Recent trends in electoral geography', *Proceedings of the New Zealand Geography Conference*, 1: 7–12.

Malone, K. (2002) 'Street life: youth, culture and competing uses of public space', *Environment and Urbanization*, 14: 57–168.

Mamadouh, V. (2003) 'Some notes on the politics of political geography', *Political Geography*, 22, 663–75.

Mann, M. (1984) 'The autonomous power of the state: its origins, mechanisms and results', *European Journal of Sociology*, 25: 185–213. Reprinted 1997 in J. Agnew (ed.) *Political Geography: A Reader*, London: Arnold.

Mann, M. (1986) *The Sources of Social Power* I, *The Beginning to AD 1760*, Cambridge: Cambridge University Press.

Mann, M. (1988) *States, War and Capitalism: Studies in Political Sociology*, Oxford: Blackwell.

Mann, M. (1997) 'Has globalization ended the rise and rise of the nation state?' *Review of International Political Economy*, 4: 472–96.

Mann, M. (2003) *Incoherent Empire*, London: Verso.

Markus, T.A. (1993) *Buildings and Power*, London: Routledge.

Markusen, A. (1999) 'Fuzzy concepts, scanty evidence and policy distance: the case for rigour and policy relevance in critical regional studies', *Regional Studies*, 33: 869–86.

Marquez, G.G. (1998) *One Hundred Years of Solitude*, London: Penguin.

Marshall, M. (1987) *Long Waves of Regional Development*, London: Macmillan.

Marshall, T. (1950) *Citizenship and Social Class*, Cambridge: Cambridge University Press.

Marston, S.A. (1989) 'Public rituals and community

power: St Patrick's Day parades in Lowell, Massachusetts, 1841–1874', *Political Geography Quarterly*, 8: 255–69.

Marston, S.A. (2000) 'The social construction of scale', *Progress in Human Geography*, 24: 219–42.

Marston, S.A. (2003) 'Political geography in question', *Political Geography*, 22: 633–6.

Martin, D.G. (2002) 'Constructing the "neighbourhood sphere": gender and community organizing', *Gender, Place and Culture*, 9: 333–50.

Martin, R. (2001) 'Geography and public policy: the case of the missing agenda', *Progress in Human Geography*, 25: 189–210.

Massey, D. (1984) *Spatial Divisions of Labour: Social Structures and the Geography of Production*, London: Macmillan.

Massey, D. (1991) 'The political place of locality studies', *Environment and Planning A*, 23: 267–81.

Massey, D. (1994) *Space, Place and Gender*, Cambridge: Polity Press.

Massey, D. (2000) 'Practising political relevance', *Transactions of the Institute of British Geographers*, New Series, 24: 131–4.

Massey, D. (2001) 'Geography on the agenda', *Progress in Human Geography*, 25: 5–17.

Massey, D. (2002) 'Geography, policy and politics: a response to Dorling and Shaw', *Progress in Human Geography*, 26: 645–6.

May, J. (2000) 'Of nomads and vagrants: single homelessness and narratives of home as place', *Environment and Planning D: Society and Space*, 18: 737–59.

Mead, L.M. (1997) *The New Paternalism: Supervisory Approaches to Poverty*, Washington DC: Brookings Institution Press.

Mele, C. (2000) *Selling the Lower East Side*, Minneapolis MN: University of Minnesota Press.

Mikesell, M.W. (1983) 'The myth of the nation state', *Journal of Geography*, 82: 257–60.

Miller, B. (1994) 'Political empowerment, local–central state relations, and geographically shifting political opportunity structures: strategies of the Cambridge, Massachusetts, peace movement', *Political Geography*, 13: 393–406.

Mirkovic, D. (1996) 'Ethnic conflict and genocide: reflections on ethnic cleansing in the former Yugoslavia', *Annals of the American Academy of Political and Social Science*, 548: 191–6.

Mitchell, D. (1993) 'Public housing in single-industry towns: changing landscapes of paternalism', in J. Duncan and D. Ley (eds) *Place/Culture/Representation*, London and New York: Routledge.

Mitchell, D. (2000) *Cultural Geography: A Critical Introduction*, Oxford: Blackwell.

Mohan, G. and Mohan, J. (2002) 'Placing social capital', *Progress in Human Geography*, 26: 191–210.

Moodie, A.E. (1949) *The Geography behind Politics*, London: Hutchinson.

Moore, T. (2002) 'Comments on Ron Johnston's "Manipulating maps and winning elections: measuring the impact of malapportionment and gerrymandering", *Political Geography*, 21: 33–8.

Mormont, M. (1987) 'The emergence of rural struggles and their ideological effects', *International Journal of Urban and Regional Research*, 7: 559–75.

Mormont, M. (1990) 'Who is rural? or, How to be rural: towards a sociology of the rural', in T. Marsden, P. Lowe and S. Whatmore (eds) *Rural Restructuring: Global Processes and their Responses*, London: Fulton.

Morris, J. (1968) *Pax Britannica: The Climax of an Empire*, London: Penguin.

Moulaert, F. (1996) 'Rediscovering spatial inequality in Europe: building blocks for an appropriate "regulationist" analytical framework', *Environment and Planning D: Society and Space*, 14: 155–79.

Moulaert, F. and Swyngedouw, E. (1989) 'Survey 15: a regulationist approach to the geography of flexible production systems', *Environment and Planning D: Society and Space*, 7: 327–45.

Muir, R. (1976) 'Political geography: dead duck or phoenix?' *Area*, 8: 195–200.

Muir, R. (1981) *Modern Political Geography*, second edition, London: Macmillan.

Murdoch, J. and Marsden, T. (1994) *Reconstituting Rurality*, London: UCL Press.

Murdoch, J. and Marsden, T. (1995) 'The spatialization of politics: local and national actor-spaces in environmental conflict', *Transactions of the Institute of British Geographers*, 20: 368–80.

Murphy, A. (1996) 'The sovereign state system as political–territorial ideal: historical and contemporary considerations', in T. Biersteker and C. Weber (eds) *State Sovereignty as Social Construct*, Cambridge: Cambridge University Press.

Nairn, T. (1977) *The Break-up of Britain: Crisis and Neo-nationalism*, London: New Left Books.

National Geographic (1988) *Historical Atlas of the United States*, Washington DC: National Geographic.

Newman, D. and Paasi, A. (1998) 'Fences and neighbours in the postmodern world: boundary narratives in political geography', *Progress in Human Geography*, 22: 186–207.

Newton, K. (1969) 'A critique of the pluralist model', *Acta Sociologica*, 12: 209–43.

Newton, K. (1976) *Second City Politics*, Oxford: Oxford University Press.

Norris, C. (1982) *Deconstruction: Theory and Practice*, London: Methuen.

Norris, P. (1996) 'Does television erode social capital? A reply to Putnam', *PS: Political Science and Politics*, 29: 474–80.

Norris, P. (1997) 'Anatomy of a Labour landslide', *Parliamentary Affairs*, 50: 509–32.

OECD (1999) *The Local Dimension of Welfare-to-Work: An International Survey*, Paris: OECD.

Offe, C. (1984) *The Contradictions of the Welfare State*, London: Hutchinson.

Offe, C. (1985) 'New social movements: changing boundaries of the political', *Social Research*, 52: 817–68.

Offe, C. (1987) 'Challenging the boundaries of institutional politics: social movements since the 1960s', in C.S. Maier (ed.) *Changing Boundaries of the Political*, Cambridge: Cambridge University Press.

Ogborn, M. (1992) 'Local power and state regulation in nineteenth-century Britain', *Transactions of the Institute of British Geographers*, 17: 215–26.

Ogborn, M. (1998) 'The capacities of the state: Charles Davenant and the management of the Excise, 1683–1698', *Journal of Historical Geography*, 24: 289–312.

Ohmae, K. (1996) *The End of the Nation State*, London: HarperCollins.

Olson, M. (1968) *The Logic of Collective Action*, New York: Schocken.

Orwell, G. (1949) *Nineteen Eighty-four*, London: Secker & Warburg.

Osborne, B.S. (1998) 'Constructing landscapes of power: the George Etienne Cartier monument, Canada', *Journal of Historical Geography*, 24: 431–58.

O'Tuathail, G. (1992) 'Putting Mackinder in his place', *Political Geography*, 11: 100–18.

O'Tuathail, G. (1996) *Critical Geopolitics*, London: Routledge.

Paasi, A. (1996) *Territories, Boundaries and Consciousness: The Changing Geographies of the Finnish–Russian border*, Chichester: Wiley.

Painter, J. (1995) *Politics, Geography and 'Political Geography'*, London: Arnold.

Painter, J. (2002) 'Multilevel citizenship, identity and regions in contemporary Europe', in J. Anderson (ed.) *Transnational Democracy: Political Spaces and Border Crossings*, London: Routledge.

Painter, J. (2003) 'Towards a post-disciplinary political geography', *Political Geography*, 22: 637–9.

Painter, J. and Goodwin, M. (1995) 'Local governance and concrete research: investigating the uneven development of regulation', *Economy and Society*, 24: 334–56.

Painter, J. and Goodwin, M. (2000) 'Local governance after Fordism: a regulationist perspective', in G. Stoker (ed.) *The New Politics of British Local Governance*, London: Macmillan.

Painter, J. and Philo, C. (1995) 'Spaces of citizenship', *Political Geography*, 14: 107–20.

Parker, G. (1998) *Geopolitics: Past, Present and Future*, London: Pinter.

Parker, G. (2002) *Citizenships, Contingency and the Countryside: Rights, Culture, Land and the Environment*, London: Routledge.

Paterson, J.H. (1987) 'German geopolitics reassessed', *Political Geography Quarterly*, 6: 107–14.

Pattie, C., Johnston, R., Dorling, D., Rossiter, D., Tunstall, H. and MacAllister, I. (1997) 'New Labour, new geography? The electoral geography of the 1997 British general election', *Area*, 29: 253–9.

Pawson, E. (1992) 'Two New Zealands: Maori and European', in K. Anderson and F. Gale (eds) *Inventing Places: Studies in Cultural Geography*, London: Longman.

Peck, J. (1995) 'Moving and shaking: business élites, state localism and urban privatism', *Progress in Human Geography*, 19: 16–46.

Peck, J. (1998) 'Workfare in the sun: politics, representation, and method in US welfare-to-work strategies', *Political Geography*, 17: 535–66.

Peck, J. (1999) 'Editorial. Grey geography?' *Transactions of the Institute of British Geographers*, 24: 131–5.

Peck, J. (2000) 'Jumping in, joining up and getting on', *Transactions of the Institute of British Geographers*, 25: 255–8.

Peck, J. (2001) *Workfare States*, New York: Guilford Press.

Peck, J. and Jones, M. (1995) 'Training and Enterprise Councils: Schumpeterian workfare state, or what?' *Environment and Planning A*, 27: 1361–96.

Peck, J. and Tickell, A. (1992) 'Local modes of social regulation? Regulation theory, Thatcherism and uneven development', *Geoforum*, 23: 347–63.

Peck, J. and Tickell, A. (1994) 'Searching for a new institutional fix: the after-Fordist crisis and global–local disorder', in A. Amin (ed.) *Post-Fordism: A Reader*, Oxford: Blackwell.

Peck, J. and Tickell, A. (1995) 'The social regulation of uneven development: "regulatory deficit", England's south-east, and the collapse of Thatcherism', *Environment and Planning A*, 27: 15–40.

Peet, R. and Thrift, N. (1989) 'Political economy and human geography', in R. Peet and N. Thrift (eds) *New Model in Geography: The Political Economy Perspective*, London: Unwin Hyman.

Perreault, T. (2003) 'Changing places: transnational networks, ethnic politics and community development in the Ecuadorian Amazon', *Political Geography*, 22: 61–88.

Philo, C. (1994) 'Political geography and everything: invited notes on "transpolitical geography"', *Geoforum*, 25: 525–32.

Pile, S. and Keith, M. (eds) (1997) *Geographies of Resistance*, London: Routledge.

Pilger, J. (1992) *Distant Voices*, London: Vintage.

Pincetl, S. (1994) 'Challenges to citizenship: Latino immigrants and political organizing in the Los Angeles area', *Environment and Planning A*, 26: 895–914.

Piore, M. and Sabel, C. (1984) *The Second Industrial Divide: Possibilities for Prosperity*, New York: Basic Books.

Pollard, J., Henry, N., Bryson, J. and Daniels, P. (2000) 'Shades of grey? Geographers and policy', *Transactions of the Institute of British Geographers*, 25: 243–8.

Polsby, N. (1980) *Community Power and Political Theory*, New Haven CT: Yale University Press.

Poulantzas, N. (1973) *Political Power and Social Class*, London: New Left Books.

Prescott, J.R.V. (1965) *The Geography of Frontiers and Boundaries*, London: Hutchinson.

Prescott, J.R.V. (1972) *Political Geography*, London: Methuen.

Prince, H. (1971) 'Questions of social relevance', *Area*, 3: 150–3.

Purcell, M. (2001) 'Metropolitan political reorganization as a politics of urban growth: the case of San Fernando Valley secession', *Political Geography*, 20: 613–33.

Purcell, M. (2002) 'The state, regulation, and global restructuring: reasserting the political in political economy', *Review of International Political Economy*, 9: 284–318.

Purcell, M. (2003) 'Islands of practice and the Marston/Brenner debate: toward a more synthetic critical human geography', *Progress in Human Geography*, 27: 317–32.

Putnam, R. (1993) *Making Democracy Work*, Princeton NJ: Princeton University Press.

Putnam, R. (2000) *Bowling Alone: The Collapse and Revival of American Community*, New York: Simon & Schuster.

Rabinow, P. (ed.) (1984) *The Foucault Reader*, Harmondsworth: Penguin.

Radcliffe, S. (1999) 'Embodying national identities: mestizo men and white women in Ecuadorian racial-national imaginaries', *Transactions of the Institute of British Geographers*, 24: 213–26.

Reich, R. (1992) *The Work of Nations*, New York: Vintage.

Reynolds, D. (1993) 'Political geography: closer encounters with the state, contemporary political economy, and social theory', *Progress in Human Geography*, 17: 389–403.

Reynolds, S. (1984) *Kingdoms and Communities in Western Europe, 900–1300*, Oxford: Oxford University Press.

Richardson, J. (2000) *Partnerships in Communities: Re-weaving the Fabric of Rural America*, Washington DC: Island Press.

Richardson, J. and Turok, I. (1992) *Local Economic Development: Report to the National Training Task Force*, Glasgow: Centre for Planning, University of Strathclyde.

Robbins, P. (1998) 'Authority and environment: institutional landscapes in Rajastan, India', *Annals of the Association of American Geographers*, 88: 410–35.

Robbins, P. (2003) 'Political ecology in political geography', *Political Geography*, 22: 641–5.

Robinson, J. (2003) 'Political geography in a post-colonial context', *Political Geography*, 22: 647–51.

Robson, B. (1972) 'Editorial comment. The corridors of geography', *Area*, 4: 213–14.

Robson, B. (1976) 'Houses and people in the city', *Transactions of the Institute of British Geographers*, New Series, 1: 1.

Robson, B., Bradford, M., Deas, I., Hall, E., Harrison, E., Parkinson, M., Evans, R., Garside, P. and Harding, A. (1994) *Assessing the Impact of Urban Policy*, London: HMSO.

Rokkan, S. (1970) *Citizens, Elections, Parties*, New York: McKay.

Rokkan, S. (1980) 'Territories, centres and peripheries: toward a geoethnic–geoeconomic–geopolitical model of differentiation within Western Europe', in J. Gottmann (ed.) *Centre and Periphery: Spatial Variation in Politics*, London: Sage.

Rose, G. (1988) 'Locality, politics and culture: Poplar in the 1920s', *Environment and Planning D: Society and Space*, 6: 151–68.

Rose, N. (1993) 'Government, authority and expertise in advanced liberalism', *Economy and Society*, 22: 283–99.

Rose, N. (1996) 'Governing "advanced liberal democracies"', in A. Barry, T. Osborne and N. Rose (eds) *Foucault and Political Reason*, London: UCL Press.

Routledge, P. (1992) 'Putting politics in its place: Baliapal, India, as a terrain of resistance', *Political Geography*, 11: 588–611. Reprinted in J. Agnew (ed.) *Political Geography: A Reader*, London: Arnold.

Routledge, P. (1997a) 'The imagineering of resistance: Pollock Free State and the practice of postmodern politics', *Transactions of the Institute of British Geographers*, 22: 359–76.

Routledge, P. (1997b) 'A spatiality of resistances: theory and practice in Nepal's revolution of 1990', in S. Pile and M. Keith (eds) *Geographies of Resistance*, London: Routledge.

Routledge, P. (2003) 'Convergence space: process geographies of grassroots globalization networks', *Transactions of the Institute of British Geographers*, 28: 333–49.

Sack, R. (1983) 'Human territoriality: a theory', *Annals of the Association of American Geographers*, 73: 55–74.

Sack, R. (1986) *Human Territoriality: its Theory and History*, Cambridge: Cambridge University Press.

Said, E. (1978) *Orientalism*, London: Penguin.

Said, E. (1993) *Culture and Imperialism*, London: Penguin.

Sanders, D. (2000) 'Electoral politics after 1997', in N. Abercrombie and A. Warde (eds) *The Contemporary British Society Reader*, Cambridge: Polity Press.

Saussure, F. de (1983) *Course in General Linguistics*, trans. R. Harris, London: Duckworth.

Schwarzmantel, J. (1994) *The State in Contemporary Society: An Introduction*, London: Harvester Wheatsheaf.

Scott, A. (1988a) 'Flexible production systems and regional development: the rise of new industrial spaces in North America and Western Europe', *International Journal of Urban and Regional Research*, 12: 171–85.

Scott, A. (1988b) *New Industrial Spaces: Flexible Production Organisation and Regional Development in North America and Western Europe*, London: Pion.

Scott, J.C. (1998) *Seeing like a State: How certain Schemes to improve the Human Condition have Failed*, New Haven CT: Yale University Press.

Scottish Geographical Journal (1999) 'Relevance in human geography', *Scottish Geographical Journal* (special issue), 115: 91–165.

Shambaugh, D. (1995) *Greater China: The Next Superpower?* Oxford: Oxford University Press.

Sharp, J. (1993) 'Publishing American identity: popular geopolitics, myth and the *Reader's Digest*', *Political Geography*, 12: 491–503.

Sharp, J. (1999) 'Critical geopolitics', in P. Cloke, P. Crang and M. Goodwin (eds) *Introducing Human Geographies*, London: Arnold.

Sharp, J., Routledge, P., Philo, C. and Paddison, R. (eds) (2000) *Entanglements of Power: Geographies of Domination/Resistance*, London: Routledge.

Shields, R. (1991) *Places on the Margin: Alternative Geographies of Modernity*, London: Routledge.

Short, J.R. (1991) *Imagined Country: Society, Culture and Environment*, London: Routledge.

Sidorov, D. (2000) 'National monumentalization and the politics of scale: the resurrections of the cathedral of Christ the Savior in Moscow', *Annals of the Association of American Geographers*, 90: 548–72.

Sismondo, S. (1993) 'Some social constructions', *Social Science Studies*, 23: 515–53.

Smith, A.D. (1986) *The Ethnic Origins of Nations*, Oxford: Blackwell.

Smith, A.D. (1991) *National Identity*, London: Penguin.

Smith, A.D. (1996) 'Memory and modernity: reflections on Ernest Gellner's theory of nationalism', *Nations and Nationalism*, 2: 371–88.

Smith, A.D. (1998) *Nationalism and Modernism: A Critical Survey of Recent Theories of Nations and Nationalism*, London: Routledge.

Smith, D. (1971) 'Radical geography – the next revolution?' *Area*, 3: 153–7.

Smith, F.M. (2000) 'The neighbourhood as a site for contesting German reunification', in J. Sharp, P. Routledge, C. Philo and R. Paddison (eds) *Entanglements of Power*, London: Routledge.

Smith, G. (1989) 'Privilege and place in Soviet society', in D. Gregory and R. Walford (eds) *Horizons in Human Geography*, Basingstoke: Macmillan.

Smith, G. (1996) 'The Soviet state and nationalities policy', in G. Smith (ed.) *The Nationalities Question in the Post-Soviet States*, London: Longman.

Smith, N. (1987) 'Dangers of the empirical return: some comments on the CURS initiative', *Antipode*, 19: 397–406.

Smith, N. (1992) 'Contours of a spatialized politics: homeless vehicles and the production of geographical scale', *Social Text*, 33: 55–81.

Smith, N. (1996) *The New Urban Frontier: Gentrification and the Revanchist City*, London and New York: Routledge.

Smith, N. (2002) 'Scales of terror: the manufacturing of nationalism and the war for US globalism', in M. Sorkin and S. Zukin (eds) *After the World Trade Center*, New York: Routledge.

Smith, N. (2003) 'Remaking scale: competition and co-operation in pre-national and post-national Europe', in N. Brenner, B. Jessop, M. Jones and G. McLeod (eds) (2003), *State/Space: A Reader*, Oxford: Blackwell, pp. 227–38.

Smith, S.J. (1989) 'Society, space and citizenship: a human geography for the "new times"?' *Transactions of the Institute of British Geographers*, 14: 144–56.

Smith, S.J. (1993) 'Bounding the Borders: claiming space and making place in rural Scotland', *Transactions of the Institute of British Geographers*, 18: 291–308.

Soja, E. (1968) 'Communications and territorial integration in East Africa', *East Lakes Geographer*, 4: 39–57.

Sorkin, M. and Zukin, S. (eds) (2002) *After the World Trade Center*, New York: Routledge.

Spruyt, H. (1994) *The Sovereign State and its Competitors: An Analysis of Systems Change*, Princeton NJ: Princeton University Press.

Spykman, N.J. (1942) *America's Strategy in World Politics*, New York: Harcourt Brace.

Spykman, N.J. (1944) *The Geography of the Peace*, New York: Harcourt Brace.

Stiell, B. and England, K. (1997) 'Domestic distinctions: constructing difference among paid domestic workers in Toronto', *Gender, Place and Culture*, 4: 339–59.

Stoler, A.L. (1991) 'Carnal knowledge and imperial

power: gender, race, and morality in colonial Asia', in M. di Leonardo (ed.) *Gender at the Crossroads of Knowledge: Feminist Anthropology in the Postmodern Era*, Los Angeles: University of California Press.

Stone, C. (1988) 'Pre-emptive power: Floyd Hunter's *Community Power Structure* reconsidered', *American Journal of Political Science*, 32: 82–104.

Stone, C. (1989) *Regime Politics: Governing Atlanta, 1946–1988*, Lawrence KS: University of Kansas Press.

Storper, M. and Scott, A. (eds) (1992) *Pathways to Industrialization and Regional Development*, London: Routledge.

Sui, D.Z. and Hugill, P.J. (2002) 'A GIS-based spatial analysis on neighborhood effects and voter turn-out: a case study of College Station, Texas', *Political Geography*, 21: 159–73.

Swyngedouw, E. (1997) 'Neither global nor local: "glocalization" and the politics of scale', in K. Cox (ed.) *Spaces of Globalization*, New York: Guilford Press.

Taylor, P.J. (1985) *Political Geography: World Economy, Nation State and Locality*, London: Longman.

Taylor, P.J. (1988) 'World systems analysis and regional geography', *Professional Geographer*, 40: 259–65.

Taylor, P.J. (1994a) 'Political geography', in R.J. Johnston, D. Gregory and D. M. Smith (eds) *The Dictionary of Human Geography*, third edition, Oxford: Blackwell.

Taylor, P.J. (1994b) 'The state as container: territoriality in the modern world-system', *Progress in Human Geography*, 18: 151–62.

Taylor, P.J. (1995) 'Beyond containers: internationality, interstateness, interterritoriality', *Progress in Human Geography*, 19: 1–15.

Taylor, P.J. (2000) 'Political geography', in R.J. Johnston, D. Gregory, G. Pratt and M. Watts (eds) *The Dictionary of Human Geography*, fourth edition, Oxford: Blackwell.

Taylor, P.J. and Flint, C. (2000) *Political Geography: World Economy, Nation State and Locality*, fourth edition, London: Prentice Hall.

Taylor, P.J. and Johnston, R.J. (1979) *Geography of Elections*, London: Penguin.

Terzani, T. (1993) *Goodnight, Mister Lenin*, London: Picador.

Tickell, A. (1995) 'Reflections on "activism in the academy"', *Environment and Planning D: Society and Space*, 13: 235–7.

Tickell, A. and Peck J. (1992) 'Accumulation, regulation and the geographies of post-Fordism: missing links in regulationist research', *Progress in Human Geography*, 16: 190–218.

Tickell, A. and Peck J. (1995) 'Social regulation after Fordism: regulation theory, neo-liberalism and the global–local nexus', *Economy and Society*, 24: 357–86.

Tickell, A. and Peck, J. (1996) 'The return of the Manchester men: men's words and men's deeds in the remaking of the local state', *Transactions of the Institute of British Geographers*, 21: 595–616.

Tilly, C. (1975) 'Reflections on the history of European state-making', in C. Tilly (ed.) *The Formation of Nation States in Western Europe*, Princeton NJ: Princeton University Press.

Tilly, C. (1990) *Coercion, Capital and European States, AD 990–1992*, Oxford: Blackwell.

Toal, G. and Shelley, F. (2004) 'Political geography: from the "long 1989" to the end of the post-Cold War peace', in G.L. Gaile and C.J. Willmott (eds) *Geography in America at the Dawn of the Twenty-first Century*, New York: Oxford University Press.

Torfing, J. (1999) 'Workfare with welfare: recent reforms of the welfare state', *Journal of European Social Policy*, 9: 5–28.

Unwin, T. (1992) *The Place of Geography*, Harlow: Longman.

Unwin, T. and Hewitt, V. (2001) 'Banknotes and national identity in Central and Eastern Europe', *Political Geography*, 20: 1005–28.

Urry, J. (1981) 'Localities, regions and social class', *International Journal of Urban and Regional Research*, 5: 455–73.

Urry, J. (1987) 'Society, space and locality', *Environment and Planning D: Society and Space*, 5: 435–44.

Urry, J. (1995) *Consuming Places*, London: Routledge.

Valentine, G. (1993) 'Negotiating and managing multiple sexual identities: lesbian time–space strategies', *Transactions of the Institute of British Geographers*, 18: 237–48.

Valler, D., Wood, A., and North, P. (2000) 'Local governance and local business interests: a critical review', *Progress in Human Geography*, 24: 409–28.

Vandergeest, P. and Peluso, N. (1995) 'Territorialization and state power in Thailand', *Theory and Society*, 24: 385–426.

Walker, R. (1993) *Inside/Outside*, Cambridge: Cambridge University Press.

Wall, D. (1999) *Earth First! and the Anti-roads Movement*, London and New York: Routledge.

Wallerstein, I. (1974) *The Modern World System I, Capitalist Agriculture and the Origins of the European World Economy in the Sixteenth Century*, New York: Academic Press.

Wallerstein, I. (1979) *The Capitalist World Economy*, Cambridge: Cambridge University Press.

Wallerstein, I. (1980) *The Modern World System II, Mercantilism and the Consolidation of the European World Economy, 1600–1750*, New York: Academic Press.

Wallerstein, I. (1989) *The Modern World System III, The Second Era of Great Expansion of the Capitalist World Economy, 1730–1840*, New York: Academic Press.

Wallerstein, I. (1991) *Unthinking Social Science*, Cambridge: Polity Press.

Walsh, F. (1979) 'Time-lag in political geography', *Area*, 11: 91–2.

Walzer, N. and Jacobs, B. (eds) (1998) *Public–Private Partnerships for Local Economic Development*, Westport CT: Praeger.

Warde, A. (1989) 'Recipes for pudding: a comment on locality', *Antipode*, 21: 274–81.

Warf, B. and Waddell, C. (2002) 'Florida in the 2000 presidential election: historical precedents and contemporary landscapes', *Political Geography*, 21: 85–90.

Watts, M. (1997) 'Black gold, white heat: state violence, local resistance and the national question in Nigeria', in S. Pile and M. Keith (eds) *Geographies of Resistance*, London: Routledge.

Watts, M. (2000) 'Colonialism', in R.J. Johnston, D. Gregory, G. Pratt and M. Watts (eds) *The Dictionary of Human Geography*, Oxford: Blackwell.

Watts, M. (2001) '1968 and all that . . .', *Progress in Human Geography*, 25: 157–88.

Weber, E. (1977) *Peasants into Frenchmen: The Modernization of Rural France, 1870–1914*, London: Chatto & Windus.

Webber, M.J. (1991) 'The contemporary transition', *Environment and Planning D: Society and Space*, 9: 165–82.

Webster, G.R. and Leib, J.L. (2001) 'Whose South is it anyway? Race and the Confederate battle flag in South Carolina', *Political Geography*, 20: 271–300.

Weigert, H. (ed.) (1949) *New Compass of the World: A Symposium in Political Geography*, London: Harrap.

Weizman, E. (2002) *The Politics of Verticality*, online. Available http://www.opendemocracy.net (12 September 2003).

Western, J. (1996) *Outcast Cape Town*, Berkeley CA: University of California Press.

Whelan, Y. (2002) 'The construction and destruction of a colonial landscape: monuments to British monarchs in Dublin before and after independence', *Journal of Historical Geography*, 28: 508–33.

Williams, C.H. (ed.) (1993) *Political Geographies of the New World Order*, London: Belhaven Press.

Williams, C.H. and Smith, A.D. (1983) 'The national construction of social space', *Progress in Human Geography*, 7: 502–18.

Willis, C. (1995) *Form follows Finance: Skyscrapers and Skylines in New York and Chicago*, New York: Princeton Architectural Press.

Wolfinger, R.E. (1971) 'Nondecisions and the study of local politics', *American Political Science Review*, 65: 1063–80.

Wood, A. (1998) 'Making sense of entrepreneurialism', *Scottish Geographical Magazine*, 114: 120–3.

Wood, W.B. (2001) 'Geographic aspects of genocide: a comparison of Bosnia and Rwanda', *Transactions of the Institute of British Geographers*, 26: 57–76.

Woods, M. (1997) 'Discourses of power and rurality: local politics in Somerset in the twentieth century', *Political Geography*, 16: 453–78.

Woods, M. (1998a) 'Rethinking elites: networks, space and local politics', *Environment and Planning A*, 30: 2101–19.

Woods, M. (1998b) 'Advocating rurality? The

repositioning of rural local government', *Journal of Rural Studies*, 14: 13–26.

Woods, M. (1999) 'Performing power: local politics and the Taunton pageant of 1928', *Journal of Historical Geography*, 25: 57–74.

Woods, M. (2003a) 'Deconstructing rural protest: the emergence of a new social movement', *Journal of Rural Studies*, 19: 309–25.

Woods, M. (2003b) 'Politics and protest in the contemporary countryside', in M. Kneafsey and L. Holloway (eds) *Geographies of Rural Cultures and Societies*, London: Ashgate.

Woods, M. (2004) 'Political articulation: the modalities of new critical politics of rural citizenship', in P. Cloke, T. Marsden and P. Mooney (eds) *The Handbook of Rural Studies*, London: Sage.

Yanow, D. (1995) 'Built space as story: the policy stories that buildings tell', *Policy Studies Journal*, 23: 407–422.

Yeung, H.W-C. (1998) 'Capital, state and space: contesting the borderless world', *Transactions of the Institute of British Geographers*, 23: 291–309.

Yuval-Davis, N. (1999) 'The multi-layered citizen: citizenship at the age of "glocalization"', *International Feminist Journal of Politics*, 1: 119–36.

Yuval-Davis, N. (2000) 'Citizenship, territoriality, and the gendered construction of difference', in E. Isin (ed.) *Democracy, Citizenship and the Global City*, London: Routledge.

Zukin, S. (1991) *Landscapes of Power: From Detroit to Disneyworld*, Berkeley CA: University of California Press.

Index